Second International Symposium on

PHYSICS AND APPLICATIONS OF AMORPHOUS SEMICONDUCTORS

Second International Workshop on

PHYSICS AND APPLICATIONS OF AMORPHOUS SEMICONDUCTORS

— Optoelectronic and Photovoltaic Devices —

Torino, Italy   September 12–16, 1988

Edited by **I.S.I.** — **F Demichelis**
Politechnico di Torino

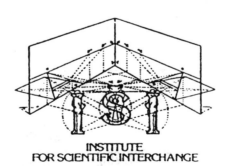

INSTITUTE
FOR SCIENTIFIC INTERCHANGE

World Scientific
Singapore • New Jersey • London • Hong Kong

Published by
World Scientific Publishing Co. Pte. Ltd.,
P O Box 128, Farrer Road, Singapore 9128
USA office: 687 Hartwell Street, Teaneck, NJ 07666
UK office: 73 Lynton Mead, Totteridge, London N20 8DH

Library of Congress Cataloging-in-Publication data is available.

Second International Symposium on
PHYSICS AND APPLICATIONS OF AMORPHOUS SEMICONDUCTORS

Copyright © 1990 by World Scientific Publishing Co. Pte. Ltd.

All rights reserved. This book, or parts thereof, may not be reproduced in any form or by any means, electronic or mechanical, including photocopying, recording or any information storage and retrieval system now known or to be invented, without written permission from the Publisher.

ISBN 9971-50-879-6

Printed in Singapore by Loi Printing Pte. Ltd.

PREFACE

This volume contains papers presented at the Second International Workshop on Physics and Applications of Amorphous Semiconductors, which was held in September 1988 at the Institute for Scientific Interchange (ISI). In consideration of the success of the 1st Workshop held last year, the ISI has kindly favoured the repetition of the Conference.

This second Workshop deals with two fundamental applications of amorphous semiconductors in the most recent technologies: photovoltaic and optoelectronic devices.

The papers cover the basic processes which occur in amorphous semiconductors, new semiconducting materials and their applications in the two specific fields.

In Part I, the presentations dealing with fundamental concepts and related to new materials are collected; Part II contains papers concerning applications.

I express my sincere thanks to the following institutions, that, by their financial support, enabled the organization of the Workshop. First of all to the ISI and to the Politecnico of Turin.

Many thanks are due to the Enea, the Centro Ricerche Fiat, the Societá Italiana Vetro, the Cassa di Risparmio di Torino, the Banca San Paolo di Torino too.

I am indebted to the members of the Scientific Committee for their suggestions and contributions. I am particularly grateful to the Scientific Organizers for their valuable help during the Workshop and for their editorial assistance in the preparation of this Volume.

THE GENERAL CHAIRMAN
Prof. F. DEMICHELIS

Second International Symposium on
PHYSICS AND APPLICATIONS OF AMORPHOUS SEMICONDUCTORS

SCIENTIFIC COMMITTEE

F. Demichelis (Chairman)	Turin	−Italy
E. Economou	Crete	−Greece
Y. Hamakawa	Osaka	−Japan
A. Madan	Wheat Ridge	−USA
T. Moustakas	Boston	−USA
I. Solomon	Palaiseau	−France
Y. Tanaka	Tokio	−Japan
C. Wronski	Philadelphia	−USA

ORGANIZING COMMITTEE

G. Kaniadakis	Turin	−Italy
A. Tagliaferro	Turin	−Italy
E. Tresso	Turin	−Italy

CONTENTS

Preface v

PART I
FUNDAMENTALS

Theories of Disorder: From Microscopic Properties to Macroscopic Phenomena 2
   Y. Bar-Yam, F. R. Shapiro, Xiaomei Wang & J. D. Joannopoulos

Defects in Disordered Systems 18
   Y. Bar-Yam

A Comparison of the Optical Absorption Edge of Crystalline and Amorphous Silicon 28
   G. Cody

Control of Reaction on Substrate for Propagation of Si-Network 63
   I. Shimizu

Changes in Gap-State Profiles of P-doped a-Si:H Induced by Light Soaking and Thermal Quenching 77
   H. Okushi, T. Furui, R. Banerjee & K. Tanaka

Subband Optical Transition in Amorphous Silicon Quantum Wells Systems 92
   K. Hattori, H. Okamoto & Y. Hamakawa

Growth of Polycrystalline Diamond Films by Filament Assisted CVD of Hydrocarbons 108
   T. D. Moustakas & R. G. Buckley

Group IV Elements in Amorphous Semiconductor Alloys 128
   F. Demichelis & A. Tagliaferro

Annealing Effects on Properties of a-CSiGe:H Alloys 136
   G. Kaniadakis, C. F. Pirri & E. Tresso

PART II
DEVICES

Transport in Schottky Barrier Structures on Amorphous Semiconductors — 150
S. J. Fonash & P. J. McElheny

Generations and Recombinations in a-Si in Time Variable Conditions — 167
J. Furlan

Computer Modelling of Internal and External Properties of a-Si:H Solar Cells — 175
F. Smole

A Non-Photovoltaic Application of Amorphous Silicon: Electroluminescent Display with Pixel Memory — 183
I. Solomon, P. Thioulouse & M. Hallerdt

Application of a-Si Diodes to Liquid Crystal Display — 195
K. Urabe, M. Kamiyama, E. Tanabe & H. Sakai

Possible Applications of Amorphous Silicon in Nuclear and X-Ray Physics — 205
C. Manfredotti

Auger and Electrical Study of the TCO/Si Interface in Amorphous Silicon Devices — 232
G. Grillo, G. Conte, D. Della Sala, F. Galluzzi & V. Vittori

PART I
FUNDAMENTALS

THEORIES OF DISORDER:

FROM MICROSCOPIC PROPERTIES TO MACROSCOPIC PHENOMENA

Y. Bar-Yam,[a,c] F. R. Shapiro,[b,c] Xiaomei Wang,[c] and J. D. Joannopoulos[c]

[a] Materials Research Dept., Weizmann Institute of Science, Rehovot 76100, Israel
[b] Dept. of Electrical and Computer Engineering, Drexel University, Philadelphia, PA 19104
[c] MIT Center for Materials Science and Engineering, Cambridge, MA 02139 USA

Experimental properties of amorphous semiconductors seem to be a collection of unrelated and complex phenomena. Models and theories tend to describe one property ignoring all others and have limited predictive value. A thermodynamic ensemble theory has connected many of the observed properties of disordered systems. Microscopic simulations then connect complex macroscopic behavior to microscopic properties. Predictions have been described for the properties of defects, band tails (Urbach edge), doping, the freezing-in transition (including the glass transition), effective ground states, metastable changes (Staebler-Wronski effect), and alloys. Here we extend these results to dc conductivity (Meyer-Neldel Rule), transient response, and dispersive transport. We emphasize the predictive power of this theoretical approach and applications to technological problems. We find excellent quantitative agreement of our simulations of the dc conductivity as a function of temperature and doping level, and the Meyer-Neldel rule is found to hold for a wide range of relevant parameters. Our simulations of transient response make explicit the connection of theory, experiment and microscopic material properties. We emphasize the development of a quantitative understanding of the microscopic properties reflected in transient response. Taken together, these efforts make progress in demystifying the phenomenology of disordered solids.

I. INTRODUCTION

In nature, disordered materials and crystals often have quite similar properties. Quartz and glass are both transparent insulators. Other properties of disordered materials are quite different. Much of our understanding of materials is based on theories which rely on order as a central component. Indeed, from a textbook in introductory condensed matter physics we would be led to believe that crystals are the generic form of matter. Moreover, the theoretical discussion of material properties begins by assuming periodicity. The objective of theories of disordered materials is thus to develop new tools for understanding materials which can describe both ordered and disordered systems and describe correctly both the similarities and differences. While many disordered materials are not in equilibrium, we have found that a thermodynamic ensemble theory is able to explain many features of disordered materials relating to electronic and structural properties, statics and dynamics.[1]

In order to establish a connection to the experimental observations, general properties can be derived which are independent of the specifics of the material. More specific properties require an understanding of material specific microstructure which may be obtained from *ab-initio* calculations or by comparison with experiment. The ability of the theory to consistently describe a large amount of seemingly and otherwise unrelated experimental information is a powerful argument for its validity. New predictions of unmeasured experimental phenomena are yet to be confirmed and would enable its verification.

For many properties of disordered materials, the connection between macroscopic observations and microscopic material properties presents an essential obstacle for testing theories, understanding experiments, and, for technological applications, predicting the behavior of new devices based on existing measurements or theory. In this work we extend our analysis of the behavior of disordered systems to describe the electron dc conductivity and transient response at finite temperatures.

II. DC CONDUCTIVITY AND THE MEYER-NELDEL RULE.[2]

Disorder in amorphous semiconductors results in unusual properties of the dc conductivity. We demonstrate a quantitative description of the temperature dependence of conductivity in a-Si:H. The universal activation energy dependence of the conductivity prefactor (the Meyer-Neldel rule) is reproduced. Excellent agreement with experimental results is obtained by describing disorder and defects using the general thermodynamic ensemble theory for the structure of disordered systems.

Unlike crystalline semiconductors, for which the Arrhenius plot is a straight line, n-type doped a-Si:H shows a kink at a temperature around 400 K.[3] Just below the kink temperature, the Arrhenius plot possesses a concavity which was most clearly indicated in recent experiments.[4] We display a typical experimental plot[3] in Fig. 1. While possessing these unique features, a-Si:H also

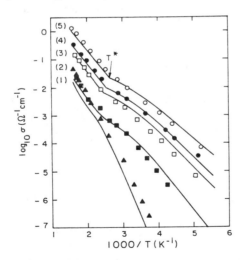

Figure 1. Comparison of the experimental (Ref. 3) and theoretical Arrhenius plot for various doping levels of phosphorus-doped a-Si:H. (1) 1 ppm, (2) 3 ppm, (3) 250 ppm, (4) 1000 ppm, and (5) 10000 ppm.

exhibits a more important universal property of the conductivity, known as the Meyer-Neldel rule[5]

$$\sigma_0 = \sigma_{00} \, e^{E_a/k T_0}. \tag{1}$$

Here, σ_0 is the extrapolation of the Arrhenius plot from room temperature to $T=\infty$. The associated slope defines the activation energy E_a. σ_{00} and T_0 are constants for a given material. This relation unifies the behavior of different samples prepared under different conditions and has been universally observed in various kinds of disordered systems, inhomogeneous semiconductors, and organic semi-insulators.[6]

A number of theoretical models have been proposed to explain the interesting properties of conductivity. For example, the kink for n-type a-Si:H has been described within the two-path conduction model[7] and the compensation model.[8] More recent detailed experiments[4] have attributed it to a transition of structural equilibrium above the kink temperature. The Meyer-Neldel rule has been investigated in the thermally assisted tunneling model.[9] The idea of a disorder-induced shift of the Fermi energy[10,11] has also been suggested. However, to date no theory has successfully given a quantitative description which is in agreement with measurements.

We use the general thermodynamic-ensemble theory for disordered systems[1] to model the electronic structure, and use the extended state conduction model to describe the transport of carriers in a-Si:H. We provide a quantitative analysis of the properties of the Fermi energy, the temperature dependence of dc conductivity, and the Meyer-Neldel rule. The theoretical results are in excellent agreement with experiments. In the temperature regime we are interested in, electronic transport in amorphous solids is likely to be dominated by extended states. Therefore, the Greenwood formula[12] provides

$$\sigma(T) = \int \mu_e(\varepsilon,T) \, g(\varepsilon,T) \, f(\varepsilon,T) \, d\varepsilon \tag{2}$$

where $\mu_e(\varepsilon,T)$, $g(\varepsilon,T)$, and $f(\varepsilon,T)$ are the mobility, density of states, and Fermi-Dirac distribution function, respectively. We further assume that the mobility $\mu_e(\varepsilon,T)$ is a constant μ_{e0} for all extended states $\varepsilon > \varepsilon_c$, and that it vanishes for $\varepsilon < \varepsilon_c$ (ε_c is the conduction mobility edge). Thus, we focus on the properties of the electronic density of states and the Fermi energy.

The central point of the thermodynamic-ensemble theory is that the amorphous solid structure itself is determined by the formation free energy of deviations from the effective ground state - an ideally bonded network. This theory assumes that there exists a freezing temperature T^*, above which both structural and electronic equilibrium can be reached, but below which the structure is frozen. Thus, when $T > T^*$, the defects with formation energy F_d have the number density $N_d = N_0 \exp(-F_d/kT)$. N_0 is the associated atomic density. When $T < T^*$, the total defect number of one type is fixed, e.g., for a defect which can exist in three charge states, $N_d = N_d^+ + N_d^0 + N_d^-$ is fixed. The formation energy of a defect in a charged state is Fermi energy dependent. As the Fermi energy μ rises the negative (positive) defects become lower (higher) in energy. For a three-charge-state defect,

$$F_d^- = F_d^0 + \varepsilon_d(0/-) - \mu$$

$$F_d^+ = F_d^0 - \varepsilon_d(+/0) + \mu$$

while for a two-charge-state phosphorous impurity

$$F_p^+ = F_p^0 - \varepsilon_p(+/0) + \mu$$

Here, $\varepsilon_d(0/-)$, $\varepsilon_d(+/0)$, and $\varepsilon_p(+/0)$ represent thermodynamic transition energies defined as the position of the Fermi energy at which the defect energies in the two different charge states are the same. The effective correlation energy is determined by $U=\varepsilon_d(0/-)-\varepsilon_d(+/0)$. In principle, the freezing temperature has a complex dependence on the properties of the defects as well as the experimental conditions. For different types of defects, T^* is not necessarily the same.

Applied to a-Si:H, this model has qualitatively explained the properties of band tails, dangling-bond defect states, and the doping dependence of the Fermi energy.[1] For the dc conductivity calculation, we again consider bands, band tails, intrinsic three-charge-state defects, and phosphorus dopants. We assume zero correlation energy and one single freezing temperature for both intrinsic defects and phosphorus defects.

The freezing temperature varies with doping. The change of T^* with doping can be derived as follows. T^* is just proportional to the activation energy for equilibration[1]

$$kT^* = -F_{eq}/\ln(N^*/A_{eq}\tau) = c_{eq} F_{eq}$$

where F_{eq} and A_{eq} are the activation energy and prefactor of equilibration and τ is an experimental time which may also depend but only weakly on T^* and F_{eq} (see Ref. 1 eqs. 1 and 2 for more details). For doped samples of a-Si:H[1] the equilibration is dominated by negatively charged species so F_{eq} is changed by the shift of the Fermi energy at T^* as

$$F_{eq} = F_{eq}(\mu_0) + q(\mu-\mu_0)$$

$$(\mu-\mu_0) = (kT^*/2) \ln ((n_0(P^+) + n_0(+)) / n_0(-))$$

where q is the charge state (-1) and μ_0 is the Fermi energy at T^* for undoped material. The expression for the Fermi energy shift $(\mu-\mu_0)$ is the thermodynamic shift[1] where $n_0(+)=n_0(-)$ is the density of positive and negative charge in the undoped material and $n_0(P^+)$ is the *initial* doping efficiency times the dopant concentration. In a-Si:H, to the desired accuracy, $n_0(P^+) \approx n(P)$ and $n_0(+)$ is given by the defect concentration in undoped material $\sim 10^{-6}$. Considering corrections only to first order, the shift in T^* is given by the expression:

$$kT^* = c_{eq} F_{eq}(\mu_0) - c_{eq} (kT^*/2) \ln (n(P) \times 10^6 + 1)$$

or

$$T^* = T_0^* - a \log_{10}[N(P)+1],$$

where $N(P)=n(P) \times 10^6$ measures the doping level in ppm. T_0^*, the undoped freezing temperature, and a are deduced from experimental data.[3,4] Note that an estimate of $F_{eq}(\mu_0)$ can be found by dividing T_0^* by a. The result ≈ 1.3 eV is in excellent agreement with other independent estimates (see Ref. 1, p. 317).

The electronic density of states of the bands can be conveniently chosen to have the Tauc form

$$g_c(\varepsilon) = A_c (\varepsilon - \varepsilon_c)^{1/2} \qquad \varepsilon > \varepsilon_c$$

$$g_v(\varepsilon) = A_v (\varepsilon_v - \varepsilon)^{1/2} \qquad \varepsilon < \varepsilon_v,$$

where $(\varepsilon_c - \varepsilon_v)$ is the Tauc optical gap. The band tails are exponential[1,13]

$$g'_c(\varepsilon) = A_c \, e^{[(\varepsilon - \varepsilon'_c)/kBT_c]}$$

$$g'_v(\varepsilon) = A_v \, e^{[-(\varepsilon - \varepsilon'_v)/kBT_v]},$$

where $T_c = T_c^0$, $T_v = T_v^0$ (for $T < T^*$), and $T_c = (T_c^0/T^*)T$, $T_v = (T_v^0/T^*)T$ for $T > T^*$. Defect states are chosen to have Gaussian distributions. Table I is the summary of input constants used in the evaluation of the density of states; the mobility gap, the Tauc optical gap, T_c^0, T_v^0, and T_0^* are chosen directly form experiments, while constant $A_c(A_v)$ is determined by the density of states at the joining point ε_c, $N_0(\varepsilon_c) \approx 3 \times 10^{21}$ eV^{-1}cm^{-3}; the peak position of the intrinsic defect density of states is determined by the Fermi energy of undoped material which is 0.6 eV below the conduction mobility edge; the peak of the phosphorus impurity level is ≈ 0.1 eV below ε_c; the formation free energies F_d^0 and F_p^0 are calculated from the measurements $n(e)/n(D^-) \approx 10^{-1}$; $n(D^-)^2/[n(p)n(si)] \approx 10^{-7}$ (see Ref. 1). In Fig. 2 we display the density of states together with the shallow states of a few typical cases in our model. It is in general agreement with the experimental data.[14]

Having a knowledge of the density of states determined by experimental data and our theory, the Fermi energy can be calculated self-consistently from charge neutrality,

$$n(e^-) + n(D^-) = n(h^+) + n(D^+) + n(P_4^+).$$

We obtain the dc conductivity by using Eq. (2) where μ_{e0} is the only fitting parameter. Finally, we approach the problem of understanding the physics behind the Meyer-Neldel rule by isolating the various components that give rise to the complexity associated with the experimental conditions. In order to draw out the general features, we consider several intrinsic physical quantities independently: doping level, freezing temperature T^*, dopant formation energy, and transition energies. Our results are presented and discussed below. We begin with the dc conductivity.

II.1 DC conductivity.

The temperature dependences of the dc conductivity for different doping levels are plotted in Fig. 1 together with experimental data. The agreement is excellent with the fitting parameter $\mu_{e0} = 50$ (cm^2/Vs). The fitted mobility is somewhat larger than the experimental value ≈ 10 (cm^2/Vs).[13] We have assumed that the atomic concentration of dopant atoms in the solid

CV Bands	$\varepsilon_c = -\varepsilon_v = 0.9$ eV, $\varepsilon'_c = -\varepsilon'_v = 0.87$ eV $A_c = A_v = 8 \times 10^{21}$ (cm^{-3}eV^{-1})
Freezing temp. (K)	$T_0^* = 440$, a=15
Band Tails (K)	$T_c^0 = 325$, $T_v^0 = 500$
Intrinsic defect	$N_0 = 5 \times 10^{22}$ (cm^{-3}eV^{-1}), $F_d^0 = 0.75$ eV ε_d peak: 0.3eV; width: 0.15 eV
Phosphorous (eV)	$F_p^0 = 0.48$, $\varepsilon(+/0)$ peak: 0.8; width 0.04

Table I: Constants describing the electronic density of states (see text).

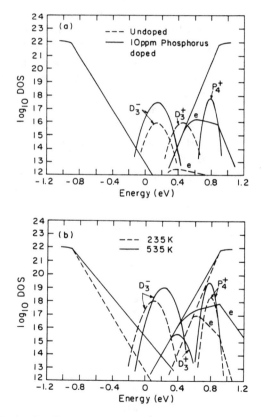

Figure 2. The density of states of a-Si:H. (a) At 300 K: dashed line for undoped samples and solid line for 10-ppm phosphorus-doped samples. (b) 1000-ppm phosphorus-doped samples: dashed line for $T < T^*$ and solid line for $T > T^*$.

phase C_{sol} is the same as that in the gas phase C_{gas}. The deviation of μ_{e0} should be reduced if we take into account the differences between C_{sol} (usually higher[15]) and C_{gas}. Since the Fermi energy plays an essential role in determining electronic transport properties, we discuss it first. For various doping levels, the overall behavior of the Fermi energy as a function of temperature (see Fig. 3) is striking. A kink is observed at T=T*. For "Impure" materials, at temperature below T*, the dominant charged defects are P_4^+ and D^-. The neutrality condition can be approximately written as $n(e^-)+n(D^-) \approx n(P_4^+)$. Since the number of D^- and P_4^+ states are fixed for T < T*, as the temperature increases the Fermi level has to decrease to satisfy the charge neutrality condition. This decrease goes faster as T goes higher.[16] For T > T*, $n(D^-)$ and $n(P_4^+)$ increase as $\exp(-F_d^-/kT)$ and $\exp(-F_p^+/kT)$, respectively. $n(P_4^+)$ increases faster than $n(D^-)$ because the formation energy of P_4^+ is smaller than that of D^- according to our model. This results in a slower decrease of the Fermi energy for T > T*. We believe this is the origin of the kink in the μ vs T curve at T = T*. [A similar behavior is found for Li-doped a-Si:H (Ref. 3).] Full understanding of the temperature dependence of the Fermi energy leads us to a clear interpretation of the conductivity behavior shown in Fig. 1. At low temperature, the Arrhenius plot is linear, because the dependence of the Fermi energy on temperature is weak. As T approaches T*, it becomes concave. The concavity corresponds to the faster drop in the Fermi energy (Fig. 3). When the temperature is above the freezing temperature, $\ln(\sigma)$ is linear with T since the Fermi level decreases linearly with T in this temperature region. The larger activation energy for T > T* (compared with that for T < T*) results from the smaller statistical shift of the Fermi level (compared with that for T < T*).

We conclude that the temperature dependence of the Fermi energy completely accounts for the properties of the dc conductivity. In particular, the discontinuity of the Fermi energy which arises because of changes in structural relaxation processes is responsible for the kink that appears in the Arrhenius plot.

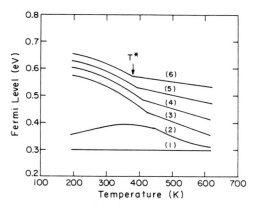

Figure 3. The Fermi energy as a function of temperature for different doping levels (1) undoped, (2) 1 ppm, (3) 10 ppm, (4) 100 ppm, (5) 1000 ppm, and (6) 10000 ppm.

II.2 Meyer-Neldel rule.

Motivated by the results just described, we next study the physics behind the Meyer-Neldel rule by describing directly the effects of varying dopants or materials processing on the dc conductivity. We present the theoretical and experimental[17] results of σ_0 as a function of the activation energy in Fig. 4. The extrapolation to obtain σ_0 is carried out by fitting the computer data of the Arrhenius plot linearly in the range of 220-320 K. The theoretical results are an accumulation of calculated data obtained by varying independently doping level, transition energies, formation free energies, and freezing temperature. In the plot, we do not explicitly distinguish between the variety of theoretical data as we are only concerned with their general behavior. However, we discuss a few typical cases below. When the doping level is varied systematically from 0.1 to 10000 ppm, the activation energy decreases from 0.62 to 0.18 eV and the scattered data fall along the line given by Eq. (1) with $T_0 \approx 800$ k, $\sigma_{00} = 1$ $\Omega^{-1}\text{cm}^{-1}$. However, the data do not spread over the entire range uniformly. A cluster locates in the region of $E_a \approx 0.19$ to 0.35 eV, while a few points representing the lightly doped samples group around $E_a \approx 0.61$ eV. This behavior is directly related to the properties of the Fermi energy dependence on doping. Both the experiments[18] and our study show that there is a jump in the Fermi energy in a very narrow range of doping levels at low temperature, which we believe is due to the competition between the D^+ and P_4^+ states. The data for smaller transition energy of the intrinsic defect are also presented. The activation energy is in the range from

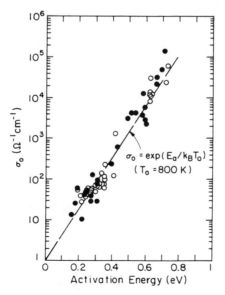

Figure 4. The preexponential factor σ_0 as a function of the activation energy. (O) theory and (●) experimental data (Ref. 17). Theoretical data are obtained by varying doping level, transition energies, formation free energies, and freezing temperature independently.

0.25 to 0.72 eV. Changing the dopant formation energy and transition energy is equivalent to changing the dopant itself in this model. The behavior is similar to that of the phosphorus dopant. More surprisingly, the data obtained by varying the freezing temperature from 350 to 520 K (this could mean varying the substrate temperature or perhaps hydrogen content in real samples) also obey the same relation. For the high dopant concentration, the higher freezing temperature produces the larger activation energy. We emphasize that μ_0 is kept constant as we change other variables.

III. SIMULATIONS OF TRANSIENT EXPERIMENTS[19]

A thorough understanding of band tails in disordered materials is essential for explaining optical and transport experiments. By simulating transient experiments, we discuss the band tail properties which are reflected in transient experiments, and the implications of such experiments for testing theories of band tails. Qualitatively and quantitatively this enables a link of theory to experiment. The results place a narrow tolerance on the precision of the exponential density of states consistent with reported experimental results.

Disorder is known to disturb the density of states (DOS) of amorphous semiconductors and insulators in surprisingly minor though significant ways in comparison to crystalline materials. Abrupt edges of the conduction and valence bands in ideal crystals are replaced by tails of localized states, which extend into the gap between the bands and cause a tail in the optical absorption. Similar tails in the optical absorption of ionic crystals at finite temperature were first seen by Urbach[20] in 1953, and have been the subject of much theoretical work [1,21–23].

In amorphous semiconductors, a significant advance was made when electrical transport experiments[24,25] suggested the presence of band tails which decrease *exponentially* toward the middle of the gap, which also explained the Urbach tails in these materials. However, a central difficulty in such experiments is in finding the bounds which they place on the band tails---the precision to which the band tail is exponential and over what energy range. Some experiments[26,27] have suggested that the band tails may be exactly exponential over an energy range as large as 0.56eV in As$_2$Se$_3$ and 0.27eV in amorphous silicon hydride (*a*-Si:H). The range of validity of the exponential behavior has potential importance for the testing of theories since some theories[22] find an exponential form for the band tails as an intermediate regime between the Halperin-Lax[21] result of $\exp[-\varepsilon^{0.5}]$ and the Gaussian form of $\exp[-\varepsilon^2]$.

Band tails are often studied by transient experiments. In these experiments, carriers are typically injected into energy states at or near the top of a band tail, and then gradually sink into states deeper in the band tail as the experiment continues. Some characteristic of the injected carriers, such as the current they can carry or their effect on the optical absorption, is measured as a function of time, and then analyzed to study the characteristics of the band tails.

As a sample experiment we simulate an electron time-of-flight (TOF) experiment on *a*–Si:H. This experiment and related ones have been widely used to study this material[28,29] and other disordered materials. In the experiment, electron-hole pairs are generated near a blocking contact at t=0, and then the electrons are forced across the 1 μm thick device by an applied voltage of 1V. The conduction band mobility μ_{e0} was set to 10 cm^2/Vs, and the hole mobility was set to 0. For a purely exponential band tail, the DOS followed the form

$$g(\varepsilon) = g_C \, e^{-(\varepsilon_C - \varepsilon)/kT_0} \qquad (3)$$

where g_C is the DOS at the conduction band edge $\varepsilon=\varepsilon_C$, and $T_0=300$K, approximately its value for the conduction band tail of a-Si:H. The states in the band tail act as trapping sites with a capture rate which is independent of energy. There is no analytic solution for the measured transient in this experiment in the presence of an arbitrary distribution of localized states.

The simulations employ a general-purpose numerical simulator for transient experiments on semiconductors and insulators[30,31]. The simulator calculates the transient as a function of time using the electron transport equation

$$J_e = en\mu_{e0}E + eD_e \frac{dn}{dx} \qquad (4)$$

and the Schockley-Read-Hall trapping equation[32]

$$\frac{dn_T}{dt} = c_n \, n \, (N_T - n_T) - e_n \, n_T \qquad (5)$$

where n_T is the number of electrons in a particular level of traps with a density N_T, n is the density of electrons in the conduction band, c_n is the electron capture rate, and

$$e_n = c_n \, g_C \, kT \, e^{-(\varepsilon_C-\varepsilon)/kT} = v_0 e^{-(\varepsilon_C-\varepsilon)/kT} \qquad (6)$$

is the emission rate as determined by detailed balance. The charge and carrier continuity equations are also used, along with Gauss's law to calculate the electric field throughout the sample. The equations are solved on a non-uniform discrete spatial grid for the specified distribution of localized states. The continuous band tails used here are modeled by 30 discrete levels at intervals of 0.014eV. Further details of the simulation are given elsewhere. [30,31] The method is more exact than other approaches which have previously been used to calculate TOF transients or particular features of them.[24,33-37] The results of this work are also applicable to TOF experiments in other materials, as well as to other experiments on carriers in band tails.

Figures 5 and 6 show the effects of adding an additional discrete level to the band tail at energy ε_r, so that the DOS is

$$g(\varepsilon) = g_C e^{-(\varepsilon_C-\varepsilon)/kT_0} + \delta(\varepsilon_C-\varepsilon_r) \, \gamma \, kT \, g_C \, e^{-(\varepsilon_C-\varepsilon_r)/kT_0} \qquad (7)$$

where γ represents the ratio of the states in the discrete level to the states within kT of ε_r in the exponential tail. The transient response of such band tails has also been studied by Silver *et al.*[38], Seynhaeve *et al.*[39], and Michiel *et al.*[35]. The case $\gamma=0$ is a purely exponential band tail, for which the current transient in Figs. 5 and 6 follows the form $I(t) \propto t^{\alpha-1}$ for t smaller than a time t_T, defined as the transit time, where $\alpha = T/T_0$. For $t>t_T$, $I(t) \propto t^{-\alpha-1}$. Both of these forms have been derived by Scher and Montroll[40] and Tiedje and Rose,[24] and their accuracy is investigated by Shapiro.[41]

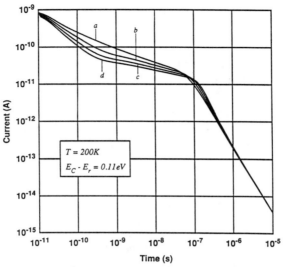

Figure 5: Simulated TOF transients with a discrete level added to the exponential band tail, with $E_C - E_r$ =0.11eV, T=200K, α = 2/3, and a) γ= 0, b) γ =0 3, c) γ=0 6, d) γ=0 9.

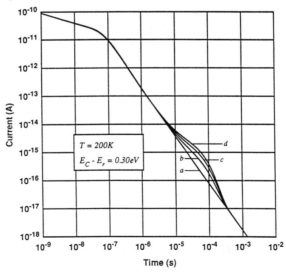

Figure 6: Simulated TOF transients with a discrete level added to the exponential band tail, with $E_C - E_r$ =0.30eV, T=200K, α = 2/3, and a) γ= 0, b) γ =0 3, c) γ=0 6, d) γ=0 9.

As suggested by Tiedje and Rose[24] and Orenstein and Kastner,[25] the primary effect of a feature in the DOS at an energy ε_r is when $e_n(\varepsilon)t \approx 1$, or

$$t \approx \frac{1}{v_0} e^{-(\varepsilon_c-\varepsilon_r)/kT} \tag{8}$$

In Figs 5 and 6, $v_0 = 8.6$ times 10^{11} s^{-1}. Therefore, when $\gamma > 0$, the second term in eq. 7 causes dips in the transients of Fig. 5 centered at about 7×10^{-10} s, and the humps in Fig. 6 centered at about 4×10^{-5} s. However, the dips and humps are quite broad, which means that details of the DOS, such as the energy range of a level or whether there are two levels with similar energies, cannot be seen in this type of experiment. A sharp feature in the density of states larger than kT $g_c e^{-(\varepsilon_c-\varepsilon)/kT_0}$ is likely to be detected, but only as a broad deviation in a transient. The effect of a change in the shape of the band tail which extends over its entire energy range can be studied by using a band tail DOS

$$g(\varepsilon) = g_c \exp\{-[(\varepsilon_c-\varepsilon)/kT_0]^m\} \tag{9}$$

This DOS is a pure exponential band tail if m=1.0. Other band tails of interest are m=2.0, the Gaussian band tail, and m=0.5, the band tail shape predicted by Halperin and Lax[21] for a 3-dimensional random material.

Figures 7 and 8 show the simulated transients for $0.8 \leq m \leq 1.0$ and $1.0 \leq m \leq 1.3$. It can be seen that the transient is very sensitive to the value of m over this narrow range. This sensitivity is most apparent in the effect of m on the transit time t_T, which is the time of the change in the slope of the transient. The value of t_T changes by nearly an order of magnitude when m changes from 1.0 to 1.1, and by nearly two orders of magnitude when m changes from 1.0 to 0.9. Many analyses of experimental data are thus seen to be very dependent on a value of m which is very close to 1.0.

Figure 9 shows current transients for m=0.5, 1.0, 1.5 and 2.0. In this figure, the forms $I(t) \propto t^{\alpha-1}$ and $I(t) \propto t^{-\alpha-1}$ are only observed for m=1.0. In the case of m=0.5, no change in slope is observed which is associated with a transit time, and a crude estimate of what the transit time should be gives $t_T \gg 10^{100}$ years. The transients with m=1.5 and m=2.0 look like transients that would be observed if there were no band tail. If one of these transients were observed in an experiment, the transit time might be used to calculate the band mobility from the formula

$$\mu_{e0} = \frac{L^2}{Vt_T} \tag{10}$$

However, this equation would give a value $\mu_{e0} = 1$ cm^2/Vs for m=2.0 and $\mu_{e0}=0.7$ cm^2/Vs for m=1.5, much smaller than the correct value $\mu_{e0}=10$ cm^2/Vs.

These results show that although time-of-flight and similar experiments are relatively insensitive to sharp features of the density of states of the band tail, they are very sensitive to changes in the form of the density of states which affect the entire band tail. This means that the experimental results of Tiedje et al. [28], Marshall et al. [29], Monroe and Kastner [26], and others clearly indicate that, except possibly for some small sharp features, band tails must have an exponential form over the range of energies to which the experiments are sensitive. For a band tail of the form in Eq. 9, the

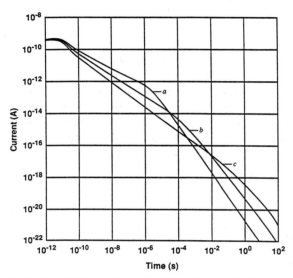

Figure 7: Simulated TOF transients at T=150K with a band tail as in eq. 9, with a) m = 1.0, b) m = 0.9, c) m = 0.8.

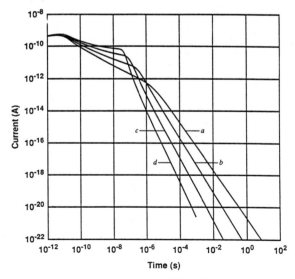

Figure 8: Simulated TOF transients at T=150K with a band tail as in eq. 9, with a) m = 1.0, b) m = 1.1, c) m = 1.2, d) m = 1.3.

value of m must be within about 0.2 of the pure exponential value of m=1.0. We further note that if Gaussian band tails are present in a material, they could escape experimental detection and yet cause a substantial error in the measured mobility.

IV. CONCLUSIONS:

In summary: a) We provide a quantitative description of the properties of the dc conductivity and Meyer-Neldel rule in a-Si:H. We find that the behavior of the Fermi energy is primarily responsible for the nature of defect states, doping, and transport. b) We have investigated quantitatively the correspondence between the electron density of states in the band tail region with the results of transient transport experiments. Simulations enable a better understanding of the experimental and theoretical work on disordered systems, demonstrating a unified understanding of material properties in these systems. The new understanding and simulation tools developed should assist in progress towards technological applications.

We gratefully acknowledge the help of Marvin Silver and the late David Adler. This work was supported in part by the U. S. National Science Foundation under Materials Research Laboratory Grant No. DMR 84-18718 by U.S. Office of Naval Research Contract No, N00014-86-K-0158, and the Solar Energy Research Institute under Contract No. DE-AC02-83CH10093. Y. Bar-Yam acknowledges assistance of the Revson Foundation and support of an Alon Fellowship.

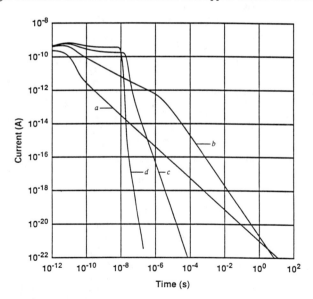

Figure 9: Simulated TOF transients at T=150K with a band tail as in eq. 9, with a) m = 0.5, b) m = 1.0, c) m = 1.5, d) m = 2.0.

REFERENCES:

1. Y. Bar-Yam, D. Adler and J. D. Joannopoulos, Proceedings of the International Symposium on Physics and Applications of Amorphous Semiconductors Torino, Sept. 1987, F. Demichelis ed. (World Scientific, 1988) p. 13; p. 317; Phys. Rev. Lett. 57, 467 (1986); in *Amorphous Silicon Semiconductors---Pure and Hydrogenated*, edited by A. Madan, M. Thompson, D. Adler, and Y. Hamakawa (Mater. Res. Soc. Proc. 95, Pittsburgh, PA 1987) p. 3
2. X. M. Wang, Y. Bar-Yam, D. Adler and J. D. Joannopoulos, Phys. Rev. B 38, 1601 (1988)
3. W. Beyer and H. Overhof, in Semiconductors and Semimetals: Hydrogenated Amorphous Silicon, Pt. C: Electronic and Transport Properties, edited by R.K. Willardson and A.C. Beer (Academic, Orlando, 1984), p.257.
4. J. Kakalios and R.A. Street, Phys. Rev. B 34, 6014 (1986); R.A. Street, J. Kakalios, and T.M. Hayes; ibid. 34, 3030 (1986).
5. W.Meyer and H. Neldel, Z. Tech. Phys. 18, 588 (1937).
6. G.G. Roberts, Phys. Rep. 60, 59 (1980).
7. Z.S. Jan, R.H. Bube, and J.C. Knights, J.Electron. Mater. 8, 47 (1979); D.A. Anderson and W.Paul, Philos. Mag. B 45, 1 (1982).
8. R.A. Street, J.Non-Cryst. Solids 77/78, 1 (1985).
9. G. Kemeny and B. Rosenberg, J.Chem. Phys. 52, 4151 (1970); 53, 3549 (1970); G.G. Roberts, J. Phys. C 4, 3167 (1971).
10. F.R. Shapiro and D. Adler, J. Non-Cryst. Solids 66, 303 (1984).
11. H. Fritzsche, Sol. Energy Mater. 3, 417 (1980).
12. O. Madelung, Introduction to Solid-State Physics (Springer-Verlag, Berlin, 1978).
13. See T. Tiedje, in Ref. 3, p. 207.
14. J. Kocka, J. Non-Cryst. Solids 90, 91 (1987).
15. For glow-discharged a-Si:H, $C_{sol} > C_{gas}$. See M. Stutzmann, D.K. Biegelsen, and R. Street, Phys. Rev.B 35, 5666 (1987).
16. We can prove that this decrease approximately goes $\approx T^2$.
17. D.E. Carlson and C.R. Wronski, in *Amorphous Semiconductors*, Topics in Applied Physics, Vol. 19. (Springer-Verlag, Berlin, 1979).
18. W.E. Spear and P.G. LeComber, Philos. Mag. 33, 935 (1976).
19. F. R. Shapiro and Y. Bar-Yam, in *Topics in Non-Crystalline Semiconductors*, edited by H. Fritzsche and A.-L. Jung, (Beijing University of Aeronautics and Astronautics, Beijing, China) 75 (1988); in *Amorphous Silicon Technology*, edited by Y. Hamakawa, P. G. LeComber, A. Madan, P. C. Taylor, and M. J. Thompson (Materials Research Society Proceedings, vol. 118, Pittsburgh, PA) 531(1988); and to be published. Simulations of double injection experiments are described in F. R. Shapiro, Y. Bar-Yam and M. Silver, IEEE Trans. El. Dev. (in press)
20. F. Urbach, Phys. Rev. 92, 1324 (1953).
21. B. I. Halperin and M. Lax, Phys. Rev. 148, 722 (1966); 153, 802 (1967)
22. S. John, C. M. Soukoulis, M. H. Cohen, and E. N. Economou, Phys. Rev. Lett. 57,1777(1986).
23. See also, Y. Toyozawa, Prog. of Theor. Phys. 20, 53 (1958); J. D. Dow and D. Redfield, Phys. Rev. Lett. 26, 762 (1971); Phys. Rev. B 5, 594 (1972); S. Abe and Y. Toyozawa, J. Phys. Soc. Jpn. 50, 2185 (1981); C. M. Soukoulis, M. H. Cohen, and E. N. Economou, Phys. Rev. Lett. 53, 616 (1984)
24. T. Tiedje and A. Rose, Sol. St. Comm. 37, 49 (1980).
25. J. Orenstein and M. Kastner, Phys. Rev. Lett. 46, 1421 (1981).
26. D. Monroe and M. A. Kastner, Phys. Rev.B 33, 8881 (1986).
27. K. Winer, I. Hirabayashi, and L. Ley, Phys. Rev. Lett. 60, 2697 (1988).
28. T. Tiedje, J. M. Cebulka, D. L. Morel, and B. Abeles, Phys. Rev. Lett. 46, 1425 (1981).
29. J. M. Marshall, R. A. Street, and M. J. Thompson, Phil. Mag. B 54, 51 (1986).
30. F. R. Shapiro and Y. Bar-Yam, J. Appl. Phys. 64, 2185 (1988).
31. F. R. Shapiro, Ph.D. thesis, MIT (1988).
32. J. C. Moll, *Physics of Semiconductors*, (McGraw-Hill, New York, 1964) pp. 91-108.
33. J. M. Marshall, H. Michiel, and G. J. Adriaenssens, Phil. Mag. B 47, 211 (1983).
34. J. M. Marshall and R. A. Street, Sol. St. Comm. 50, 91 (1984).
35. H. Michiel, G. J. Adriaenssens, and E. A. Davis, Phys. Rev. B 34, 2486 (1986).

36. J. M. Marshall, J. Berkin, and C. Main, Phil. Mag. B **56**, 641 (1987).
37. G. Seynhaeve, R. P. Barclay, and G. J. Adriaenssens, J. Non-Cryst. Sol. **97&98**,607(1987).
38. M. Silver, E. Snow, and D. Adler, Sol. St. Comm. **53**, 637 (1985); J. Non-Cryst. Sol. **77 & 78**, 455 (1985).
39. G. Seynhaeve, G. J. Adriaenssens, and H. Michiel, Sol. St. Comm. **56**, 323 (1985).
40. H. Scher and E. W. Montroll, Phys. Rev. B **12**, 2455 (1975).
41. F. R. Shapiro, Sol. St. Comm. **68**, 623 (1988).

DEFECTS IN DISORDERED SYSTEMS

Y. Bar-Yam

Materials Research Dept., Weizmann Institute for Science, Rehovot 76100, Israel
and MIT Dept. of Physics, Cambridge, MA 02139 USA

A formal discussion of the nature of defects in both crystals and disordered systems may not rely on the disruption of long range order. Short range order or topological order is an alternative concept based on bonding ideas without fundamental justification. We develop a fundamental definition of defects which can be applied to crystals and disordered systems equally. It is formally connected to the existence of experimental signatures of unique structures (phenomenological definition of defects) in materials. Our approach reveals assumptions made in a topological short range order approach and shows that topological defects are not always either well defined or to be associated with experimental defects.

I. INTRODUCTION

Defects in crystals have been conveniently defined, in the limit of zero temperature, through the disruption of long range order. This basic relationship of defects to order has inhibited the progress of understanding defects in disordered systems. For many, the concept of a defect in a disordered system is a self-contradiction.[1] In contrast, for others who have experimentally investigated disordered materials, defects are a natural concept related to the phenomenological observation of characteristic signatures similar to the defects present in crystals.[2] Conceptually these have been related to the short range order which in some materials seems to be so similar to the short range order in crystals that the defects in crystals may be thought to be directly adaptable. In most materials, the defects in a disordered system are thought to be different from those in related crystals but still associated to the breaking of a local order. The local order in these disordered materials is thought to be topological - a network structure of sites (atoms) and bonds. This order is broken by "topological" defects such as wrongly coordinated atoms.

In this article it is proposed that the phenomenological and conceptual attributes of defects are much more naturally related to an ensemble definition in which the uniqueness of the attributes of specific structures is the essential property which defines them. Order then takes a back seat, since an appropriate definition of order may always be recovered by simply defining the "order" to be the absence of such special structures. Indeed this would be precisely what we would do if confronted with a new system which had some other special low concentration unique structures - we would define them as defects and the system without them (whether attainable in principle or in practice) would then be defined as the ordered state with some appropriate measure of the order.

To illustrate this concept for defects we shall begin by discussing definitions of defects in pure crystals and extend our discussion to defects in disordered systems and specifically topological (bonding) defects.

Our approach will be thermodynamic: considering defects as created in equilibrium, we will consider the resulting ensemble.[3] The results may be applied to many disordered materials not in equilibrium through use of the effective ensemble temperature T -> T*. Since there is no obvious generalization to a fundamentally non-equilibrium ensemble we recover the approach of the sceptic for the general case: in the general case defects may indeed not always be natural objects of discussion.

This definition has many advantages, for example, it does not require that we extrapolate a crystal to absolute zero conceptually to define the defect. Furthermore it is naturally extendable to disordered systems. It seems that the definition and approach taken is not only close to our intuition but also enables a formal discussion of the properties of these systems which is theoretically lacking. We show, for example, that the hesitation of many to accept the notion of topological defects in disordered systems is non-trivial but reflects an intuitive understanding that such topological entities may indeed not always be defects. However, different kinds of topological defects may be described and have an important role to play in our understanding of defects in general within the framework developed here.

We motivate our discussion by the following comment and questions:[4] There exists experimental evidence for the existence of specific numbers of unique entities with common properties. For example, a concentration of 10^{-3} or 10^{-6} spin signals by number of atoms. We therefore ask: Why are special signatures present and what sets the scale for their number?

Our discussion begins with crystals.

II. DEFECTS IN CRYSTALS: INDEPENDENT SUBSYSTEM ASSUMPTION.

The usual thermodynamic derivation of the number of defects in a crystal begins with a consideration of the entropy of N_D defects on N crystallographic sites.[5] An equivalent derivation is simpler: Let us consider each crystal site to be essentially an independent subsystem. We then note that each possible realization of this site in thermal equilibrium will have a thermodynamic probability $e^{-\beta E}$. By integrating over all realizations that are identified as belonging to a defect (i.e. vacancy, interstitial) we define the Free energy of a defect as $F_D = -kT \ln Tr' e^{-\beta E}$ where the prime indicates tracing only over configurations identified with the defect. Then by definition of F_D we have $N_D = N e^{-\beta F_D}$. This derivation is equivalent to the usual one, but it exposes an assumption which is often taken for granted, this is the assumption that each crystallographic site is independent. The essential point here is that in a general statistical mechanical system (e.g. interacting spins) we do not usually take sites as independent, however, for a crystal we do. Our intuition tells us, for example, not to expect the crystal structure to change significantly (structural collapse) over large distances as a result of removing a single atom. This is an assumption based on experience rather than proof. Some leeway can be gained by stating that the Free energy can always be defined so that the trace takes into account local rearrangements. The simplest case would be a vacancy which relaxes to be a bridging atom between two vacancies (see Fig. 1). Even though we don't know which site to assign this defect to, the Free energy trace can be remade to enable correct counting without great difficulty. This simple case shows how we expect the basic nature of a defect to be essentially unchanged by relaxation, or, stated differently, by the interactions of the statistical mechanical system. In contrast, consider the atoms as a collection of spins, the flipping of a spin in some models would not result in a well defined object or excitation, the natural objects to define would be extended and potentially complicated.

In this sense it can be seen that our definition of defects is very closely related to the concept of elementary excitations. For elementary excitations we assume that a particular set of excitations is essentially statistically independent. In this picture, describing the formation of a vacancy, we have made the analogous assumption of independence for the subsystem consisting of one atom possibly including its local environment effectively "renormalizing" the elementary excitation. This local environment may include several or several tens of atoms.

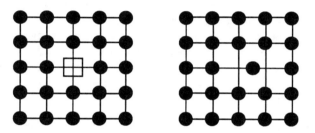

Fig. 1: Illustration of statistical independence of atom sites for the case of a vacancy in a crystal. The two illustrative minimum energy geometries show how little we expect the material structure to change as a result of removing an atom. In the second case we do not know exactly which site to assign the vacancy to, but we get the correct counting by assigning a configurational entropy of k ln 2. More drastic reconstructions could also be dealt with but at some point we would begin to call the defect something other than a vacancy.

II. DEFECTS IN OTHER STATISTICAL MECHANICAL SYSTEMS.

There are related concepts of elementary excitations which can serve as helpful analogues for our discussion of defects.

The example we consider is superfluid rotons in Helium II. The existence of a spectrum of elementary excitations whose quantum number is known with a form illustrated in Fig. 2 enables us to recognize explicitly the density of a characteristic elementary excitation at q_0 whose contribution to the system properties may be treated as due to independent particles with a density :

$$N_{Roton} = Tr' \, e^{-\beta E(\alpha)} = e^{-\beta F}$$

Where Tr' is restricted to the region around q_0. What is relevant to the analysis is that the energy rises away from the coordinate q_0. More stringently, fixing q, the restricted Free energy trace of all transverse coordinates rises away from q_0. This guarantees that for low temperatures the region of phase space around q_0 contributes to the system properties in an important way near the ground state (kT<< E_D) namely they contribute a number of identifiable unique entities to the system whose number is specified above. It is the existence of characteristic sets of entities which causes defects to be important in materials in the same sense that rotons are important to the Helium II properties.

We claim that this is a generic example of a defect in an equilibrium system because (essentially by definition) for a characteristic property to be introduced in number N this number must be related to properties (of a part of the configuration space) whose probability is larger than slightly different properties and therefore when a coordinate is properly defined the configuration space energy must always have a form as Fig. 2.

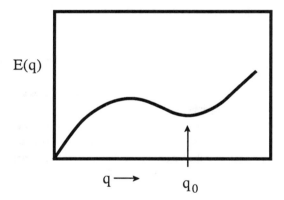

Fig. 2: Schematic energy dispersion relation for superfluid He II for temperatures below 1K. The minimum in the dispersion relation known as the Roton minimum guarantees that for low temperatures rotons are identifiable objects contributing characteristic properties to the material. Defects are similar in many ways and indeed such an energy surface should occur for defects in any system.

III. DEFECTS IN DISORDERED MATERIALS[6]

In order to generalize this discussion the first task is to define formally a defect in the ensemble sense. From the above discussion we consider systems in which statistical independence applies. To be explicit we consider "point" defects where, as for the crystal above, a small region of the material is essentially independent of its surroundings. For the crystal case this expressed itself as - in some sense - the defect causing a weak perturbation on the material structure. The analogous reverse statement is also assumed, namely, that for the likely environments the defect properties are weakly affected. So that if we allow phonons at finite temperature in a crystal, they will not completely transform the defect structure or properties. In this sense we describe the local subsystem as statistically independent. As defined, this assumption applies to high energy or dilute excitations.

It is to be emphasized at this point that statistical independence is not arbitrary. It is only valid within the material ensemble. Specifically, we do not consider a small region of material to be independent of its surroundings, we *do* assume that it is only weakly dependent on its surrounding for surroundings consistent with all but extremely rare ensemble members. So that, for example, it is known that a cluster does not have the properties of the same number of atoms in the solid, extreme conditions not generally present in the ensemble (like high pressures) affect properties arbitrarily. It is only within the ensemble that we can define statistical independence.

We now formalize this discussion.

IV. ENERGY SURFACE FOR LOCAL COORDINATES.

We consider the energy of the system $E(\{R\})$ as a function of the atomic coordinates $\{R\}$. (We treat the atomic coordinates classically, and assume the energy is obtained quantum mechanically. In general $E(\{R\}) = E(\{R\},T)$ - smooth variations as a function of T present no difficulty and phase transitions must be dealt with separately.)

We now separate the system coordinates $\{R\}$ to local coordinates $\{\alpha\}$ and all other coordinates $\{\beta\}$; where $\{\alpha\}$ are, for example, the coordinates of an atom and its nearest neighbors, or particular bond lengths and bond angles, or the lengths of smallest interconnected rings of atoms at an atom. $\{\beta\}$ are all the remaining coordinates so that $\{\alpha,\beta\}$ are completely equivalent to $\{R\}$. To define a defect we then consider the energy dependence on the local coordinates $\{\alpha\}$ allowing $\{\beta\}$ to sweep over all its values, but weighting the $\{\beta\}$ through the ensemble. The conceptual difficulty here is that the system is coupled, a choice of $\{\alpha\}$ affects the low energy configurations of $\{\beta\}$ and this might be thought to trouble us when $\{\alpha\}$ does not fully describe a defect. This does not trouble our definition however, as we are interested in identifying the energetically separated regions of phase space. Considering the ensemble averaged energy

$$<E(\{\alpha\})> = Tr_\beta \ P(\{\alpha,\beta\})E(\{\alpha,\beta\}) \ / \ Tr_\beta \ P(\{\alpha,\beta\})$$

We define as a defect each local minima in $\langle E(\{\alpha\})\rangle$ or more physically the (drainage) neighborhoods of a local minima $\{\alpha_0\}$. The global minimum, and its neighborhood is to be identified with the effective ground state. There may be several essentially equivalent minima for either ground state or defects which are to be grouped together. (note that the ground state may also be an ensemble, these are to be identified with probability of order unity ie E within kT of the global minimum.). [3,7]

Using this definition we can then describe the number of defects and the Free energy of formation:

$$N_d(\{\alpha_0\}) = N \, Tr_\beta \, Tr'_{\alpha \sim \alpha_0} P(\{\alpha,\beta\})$$

$$F_d = -kT \ln Tr_\beta \, Tr'_{\alpha \sim \alpha_0} P(\{\alpha,\beta\})$$

where the prime indicates a restricted trace over the neighborhood of $\{\alpha_0\}$, and by definition $N_d(\{\alpha_0\}) = N \, e^{-\beta F_d}$

V. COORDINATION "TOPOLOGICAL" DEFECTS.

As an example we consider a general definition of a coordination defect in a solid. This example will also serve to bring up several interesting points. Let us consider one local coordinate

$$\alpha = \sum f(|R_i - R_0|)$$

where f is a decaying function over an interatomic distance whose details are not crucial to the argument but may have to be chosen properly for a particular case. α is then a definition of the coordination of the atom at R_0. α is defined to enable the coordination to be a continuous variable rather than just a counting of neighbors within a particular radius, but to yield effectively the counting result for equidistant neighbors. We then write

$$E(\alpha) = Tr \, E(\{R\}) \, P(\{R\}) \, \delta(\alpha - \sum f(|R_i - R_0|)) \, / \, Tr \, P(\{R\}) \, \delta(\alpha - \sum f(|R_i - R_0|))$$

$$F(\alpha) = -kT \ln Tr \, P(\{R\}) \, \delta(\alpha - \sum f(|R_i - R_0|))$$

In descriptive language what we are doing is fixing α and looking at all possible atomic arrangements picking out the low energy ones (or most probable ones as weighted by $P(\{R\})$). The picture we have in mind is that as α decreases (from the global minimum value) one of the nearest neighbors is gradually distanced.[8] This distance increases until "the bond breaks". Eventually the displaced atom becomes identified with one of the second neighbors. Decreasing α still further will begin to push a second atom away. Increasing α would bring a second neighbor closer.

Bonds do not break for two atoms in isolation, the energy gradually and continuously increases as the two atoms are separated. The breaking of a bond in a solid results from structural relaxation - the rearrangements of other atoms. The central question in identifying whether a distinct defect exists is whether the energy after rising then decreases to find a new local minimum. If the energy does not decrease but continually increases then there is no special region of phase space to identify with a coordination defect. If it does decrease, then the following local minimum, as in Fig. 2, gives rise to an identifiable entity which is a defect by our definition and phenomenologically.

Our intuition tells us that in systems where bonds are "well defined", as α is lowered and an atom is distanced and before it arrives at the second shell of atoms the energy rises because there are actually two coupled constrained atoms: the central atom and the atom moving away. The atom which is gradually moving away constrains the relaxation of other atoms by its presence, and it should also be poorly coordinated. Beyond a certain point as it is further displaced it becomes decoupled, the energy begins to decrease (ie goes over a maximum and then to a new local minimum as in Fig. 2). The system resolves to a case of a lower coordination atom with the displaced atom resolving to a second neighbor in the ensemble. In such a case the defect is well defined. However, one can imagine cases in which the energy simply rises smoothly. In such a case a well defined coordination defect does not exist.[9]

We now address a subtle point. Doesn't creating a site with wrong coordination (by ±1) require at the same time the creation of another defect - so that the number of bonds is integral and the number of wrongly coordinated atoms is even?[10] This would seem to force the above expressions to yield an energy including the formation energy of both defects, rather than one as we desire. To reiterate: a problem seems to arise because we do not want the defect formation energy to include the energy of a coupled but free to move other defect. The solution of this apparent problem is, however, inherent in the above expression. In the ensemble at a finite temperature the "other" defect created may either exist or annihilate with one of the other defects in the system. Thus in addition to the defect at R_0 the members of the ensemble (with equal probability) have one more or one less defect than without the defect at R_0 (for example: total defect count is equally likely to be $<N_D>$ and $<N_D>+2$) and the trace averages these energies to get the right energy and free energy for one extra isolated defect imposed at R_0 ($<N_D>+1$ defects all together). In the ensemble sense the other wrongly coordinated atom disappears.[11]

VI. CHARGED DEFECTS.

We now consider the properties of charged defects. It is important to recognize that within an equilibrium ensemble charged and neutral defects are essentially independent. Only if the structural degrees of freedom are frozen and the electrons continue to equilibrate does the identity of a defect become separate from its charge state.[12] Indeed, for the discussion of defects in disordered systems the importance of the charge state is great because a coordination defect may not be well defined when neutral, but the corresponding charged defect may be well defined. A charged defect is defined by considering the energy of the ensemble with extra or missing electrons. Forcing the extra or missing electron to lie in the neighborhood of α we formally define a charged defect by considering the energy surface:

$$<E(q,\{\alpha\})> = \text{Tr}_\beta \ P(\{\alpha,\beta\})E(q,\{\alpha,\beta\}).$$

where $E(q,\{\alpha,\beta\})$ ranges over ensemble members with extra electrons or holes in a wavepacket at α. For each q the energy surface must have some minimum. We have the choice of labeling this minimum as an electron in the conduction band or a charged defect depending on the nature of the minimum energy configuration $\alpha_0(q)$. Several minima may exist. Implications for coordination defects are quite interesting. It is possible that in the charged energy surface a local minimum will exist corresponding to a charged defect. The neutral coordination defect may have a significantly different configuration or even be not well defined. We then may decide whether to call the charged defect a coordination defect or a self-trapped carrier.

VII. IMPURITIES AND CONSTRAINT DEFECTS:

Impurities and alloys can be directly treated in the ensemble picture with no fundamental complication. The trace over configurations β would then include also the possibility of different atom types and this would directly affect the resulting energies, free energies and probabilities of configurations. Including different atom types in $\{\alpha\}$ describes different configurations associated with the impurities (or alloy components) and thus defects associated with the impurity are included. The relevance of this to Fermi energy control ("doping") has been briefly described elsewhere.[3] The presence of impurities is most easily incorporated through the grand canonical ensemble using the chemical potential to determine the number of impurities but other choices of ensemble present no difficulty in principle.

At this point it is convenient to discuss also the concept of constrained defects. At interfaces of thin layers dislocations may be "constrained" to exist by the boundary conditions. We can make contact with this picture by imposing the constraints or adding Lagrange multipliers to establish a grand canonical ensemble picture. In the bulk of disordered materials defects have also been proposed as constrained to exist.[13] Such defects would emerge from the ensemble treatment given above. They would be distinguished from our usual concept of thermodynamic defects by the temperature dependence of the number of defects. Specifically, such constrained defects would have a given number independent of temperature and thus the formation entropy would dominate and the formation energy would be zero in the canonical ensemble and would reflect the imposition of the constraint through the relevant Lagrange multiplier (chemical potential) in the grand canonical ensemble. The essential point is that the ensemble picture described above includes defects which are related to constraints and enables one to identify them through their temperature dependence.

VIII. DEFECTS IN CRYSTALS AND AMORPHOUS NETWORKS:

We conclude by returning to the comparison of crystals and disordered materials.

The above discussion of coordination defects applies to crystals as well as disordered materials. In a crystal typically as the coordination number α is reduced, the system will switch to contain a vacancy as the lowest energy way of having a low coordination atom. The resulting vacancy might also be thought of as a system with several undercoordinated atoms (fig. 1a or 1b).

In this case the ensemble has guaranteed the coupling of the local structure to an intermediate range structure. Such effects are likely also in disordered systems. In a network model, for example, a low coordination atom even if it is not constrained by the ensemble to have other coordination defects around will have a constrained network around it. The extent to which a defect is coupled to a constrained structure depends on the variety of states in the ensemble ground state, which contribute to the structural screening.

We can discuss defects more systematically in disordered systems by considering three types of disordered systems and their coordination defects. In the first type the long range order is disrupted only by occasional ring changes. In this case local defects are likely to be the same as in the crystal and coordination defects correspond to vacancies (Materials with many crystalline polymorphs may be examples since in a glass or amorphous solid the long range order may be disrupted weakly. An isolated coordination defect would then result in long range and energetically costly strains so that vacancies would still be the dominant defect. Materials with varied stacking sequences are another example). In the second type, the ring structure is frequently disrupted at the level of single rings and the network is then flexible enough to accommodate isolated coordination defects with a constrained network surrounding it (amorphous carbon and silicon may be of this type). In the third type the network structure is sufficiently flexible to make coordination defects not exist since a continuous change in energy occurs for continuously changing coordination. Charged defects should still exist. (amorphous silicon or germanium may be of this type).[14]

IX. CONCLUSIONS.

We have proposed a definition of defects and demonstrated its connection to the phenomenological observation of defects. Our general treatement places crystals and disordered material on equal footing. This definition then gives rise to questions about our understanding of the nature of microscopic defects. Coordination "topological" defects may or may not be well defined and thus may or may not correspond to signatures observed experimentally.

Support by the Revson Foundation and an Allon Felowship is gratefully acknowledged.

REFERENCES:

[1] see P. W. Anderson, J. Phys. (Paris) 10, suppl. C4 (1976)
[2] see N. F. Mott and E. A. Davis, *Electronic Processes in Non-Crystalline Materials*, 2nd ed. (Clarendon, Oxford, 1979); N. F. Mott in *The Physics of Hydrogenated Amorphous Silicon*, J. D. Joannopoulos and G. Lucovsky, eds., Springer Verlag (1984), p. 169
[3] Y. Bar-Yam, D. Adler, and J. D. Joannopoulos, Phys. Rev. Lett. **57**, 467 (1986); M.R.S. Conf. Proc. Vol. 95, (A. Madan, M. Thompson, D. Adler and Y. Hamakawa eds.) (1987) p. 3; Proceedings of the International Symposium on Physics and Applications of Amorphous Semiconductors Torino, Sept. 1987, F. Demichelis ed. (World Scientific, 1988) p. 317
[4] Y. Bar-Yam and J. D. Joannopoulos, Proceedings of the 12th Int. Conf. on Amorphous and Liquid Semiconductors, Prague, Sept. 1987.
[5] see also Z E. Smith and S. Wagner, Phys. Rev. Lett. 59, 688 (1987); Errors in their analysis are described in Y. Bar-Yam (unpublished).
[6] For recent related work see P. C. Keliras and J. Tersoff Phys. Rev. Lett. **61**, 562 (1988)

[7] While we have not explicitly used an equilibrium Boltzmann probability this discussion relies on the existence of high and low probability regions of phase space as characterized here by their relative energies.

[8] Some care must be taken so that the decay of f does not lead to $\alpha \to 0$ too rapidly with displacing all the atoms equally. This may lead to α going to zero for a small displacement of all the atoms. Physically, in the limit $\alpha \to 0$ the energy of an isolated atom in which all atoms have been distanced far away should be higher than just displacing one atom away.

[9] It may be possible for some purposes to expand $E(\{R\})$ using effective nearest neighbor interaction energies or local coordinate energies.[6] The interaction energies are typically obtained by fitting to data representative of the systems to be looked at. The problem of understanding materials is thus mapped onto a locally interacting statistical mechanical model with Hamiltonian:

$$H = \sum E_l(\theta) = \sum E_l(|R_i - R_j|)$$

Where the latter is the simplest nearest neighbor interaction model. This is somewhat different than usual statistical mechanics problems due to the dependence of the energy on lengths and network variability but is not unheard of. The central question relevant here, is when are there defects in such a Hamiltonian and of what type. (point defects, line defects, vortices or dislocations. [note that $<E(\alpha)>$ and $E_l(\theta)$ are different being the same only for decoupled systems.]

[10] Such a constraint would not be imposed but would arise from within the energy considerations.

[11] A further point can be clarified by considering the case of vacancies and interstitials in crystalline materials. Consider the ensemble energy evaluated with an imposed vacancy at R_0. Then one might believe that an additional interstitial is also created. This interstitial however, may annihilate a vacancy or be present with equal likelihood. This would yield an ensemble formation energy of the vacancy of $1/2(E_v + E_i)$. This is not incorrect, it is the correct result for material which is dislocation free and far away from surfaces where the equilibrium between vacancies and interstitials guarantees this result (like electrons and holes in a semiconductor). If there are dislocations or surfaces in the material ensemble then the trace will separate the vacancy from the interstitial formation energy, as it should.

[12] It is worth noting that, aside from defects, a different mechanism for special signatures in a disordered material is through the electron redistribution once the material is structurally thermodynamically frozen. This is equivalent in ordered material to the signal obtained from electrons excited eg. to the conduction band. Thus, a region of phase space which is continuously connected to the ground state may have its charge state changed either by temperature changes or by excitation giving rise to a distinct material signal.

[13] J. C. Phillips, Phys. Rev. Lett. **42**, 1151 (1979); A. Ourmazd, J. C. Bean, and J. C. Phillips, Phys. Rev. Lett. **55**, 1599 (1985).

[14] It may be useful to summarize two central results which emerged from realistic studies of defects in semiconductors. First is the essential role played by the coupling of electronic state and structural energies. In almost all cases studied, electronic states are coupled to energies so that equilibrium geometries and migration paths are dramatically changed with charge state. Second, once a defect exists the bonding arrangements seem to be remarkably flexible. So much so, that we need to rethink our basic understanding of chemical bonding. [see also S. T. Pantelides Phys. Rev. Lett. 57, 2979 (1986)] This can be explained both by the effect of electronic state changes and by a concept of structural screening. Displacing a single atom results in a large energy gain. The softest modes however control the resulting relaxation leading to an energy scale characterized by the structural and electronic stiffness of the material. We expect these results to carry over to disordered systems. [An introduction to the literature can be found in Y. Bar-Yam and J. D. Joannopoulos, Proceedings of the 18th International Conf. on the Physics of Semiconductors, Stockholm, Aug. 1986, p 809; and more recent work C. G. Van de Walle, Y. Bar-Yam, and S. T. Pantelides, Phys Rev. Lett. **60**, 2761 (1988); J. Bernholc, A. Antonelli, T. M. Sel Sole, Y. Bar-Yam and S. T. Pantelides, Phys. Rev. Lett. **60**, 2689 (1988)]

A Comparison of the Optical Absorption Edge of Crystalline and Amorphous Silicon

George D. Cody
Exxon Corporate Research
Annandale, New Jersey, 08840, USA

ABSTRACT

Optical phenomena associated with the absorption edge of crystalline silicon (c-Si) and amorphous silicon hydride (a-Si:H) are presented and compared with current models and fundamental theories of the effect of disorder on the complex dielectric constant of semiconductors. Included in the discussion are the following experimental data for c-Si and a-Si:H: the magnitude and dispersion of the index of refraction; the energy dependence, dipole matrix element and density of states associated with the relevant absorption edges; the temperature dependence of the relevant optical energy gaps, and finally the magnitude and temperature dependence of the slope of the Urbach edge for each material. The comparison suggests that the optical properties of the two materials are more closely related than previously thought and that in many respects the optical properties of a-Si:H can be derived from the effect of site disorder on the electronic wave functions contributing to the the zone center ($\Gamma_{25'}$ to Γ_{15}) direct absorption edge of c-Si. A significant difference between the materials is the approximately forty percent larger band edge deformation potential for a-Si:H.

1. INTRODUCTION

1.1 The Absorption Edge of a Semiconductor

All semiconductors, whether crystalline or amorphous, exhibit an absorption edge. At the edge, the optical absorption coefficient increases by many orders of magnitude over a relatively narrow range of photon energies. This sharp increase defines the "optical energy gap" for optical transitions between the occupied valence band of the semiconductor and the unoccupied conduction band. The magnitude of the optical gap is a critical parameter for many optical applications of the semiconductor as well as an experimental test of band structure calculations. The energy dependence of the optical absorption in the vicinity of the optical gap is also of interest, since it supplies information on the details of the absorption mechanism.

There are at least three absorption edges for a semiconductor corresponding to the physics of the optical transition[1,2]:

1. <u>The direct edge of a crystalline semiconductor</u>. At this optical transition, a valence band electron is directly excited to a conduction band state of the same crystal momentum. The energy of the final state corresponds to the energy of the absorbed photon. Electron momentum is conserved in the optical transition and the final state is often an exciton. The direct gap is a prominent feature of the experimental absorption data due to its rapid variation with energy; for GaAs, the optical absorption increases from 1 cm^{-1} to 10^4 cm^{-1} over an energy range of less than tenth of a volt. The forbidden gap of the semiconductor is readily deduced from the energy of the direct edge and the exciton energy and can be directly compared to the results of band structure calculations. The low energy side of the absorption edge of direct band gap materials is often exponential over many orders of magnitude. This exponential regime is called the <u>Urbach edge</u> after its discoverer.

2. <u>The indirect edge of a crystalline semiconductor</u>. This optical edge is significantly broader than the <u>direct edge</u>; for example, at the indirect edge of crystalline silicon (c-Si) the absorption coefficient increases from 1 cm^{-1} to 10^4 cm^{-1} over 0.5 eV. A simple model for this transition assumes that the lowest energy states of the conduction band differ in crystal momenta from the highest energy states of the valence band. Since a crystalline semiconductor has long-range order, the optical transition is forbidden in first order. In second order, crystal momentum is conserved for this <u>indirect optical transition</u> by the emission or absorption of a phonon that conserves crystal momentum. The forbidden gap of the semiconductor is not an experimental feature of the absorption edge, but can be derived from changes in its derivative at energies corresponding to the absorption and emission of phonons. Again, the final electronic state can be an exciton, whose energy must be included in deriving the "forbidden gap" of the semiconductor from the <u>indirect edge</u>. Indirect gap semiconductors can also exhibit an exponential or Urbach edge on the low energy side of the indirect gap.

3. <u>The non-direct edge of amorphous semiconductors</u>. Measurement of the optical absorption of amorphous semiconductors introduces a new category of absorption edge: the

<u>non-direct edge</u>. Amorphous semiconductors exhibit an optical absorption edge that is similar to that of indirect gap semiconductors in breadth, magnitude and absence of significant features. The absence of long-range order in these semiconductors effectively eliminated first principles calculations of either the forbidden gap or the absorption process for carrier excitation above it. In the absence of fundamental theories, simple physical models were developed for the absorption edge of amorphous semiconductors[3]. The most successful of these models, the Tauc model, described the absorption edge as a <u>first order optical transition</u> involving excitation of an electron from an occupied valence band to an unoccupied conduction band with conservation of energy as the only restriction. The fit of these models to experimental optical data led to the definition of an new edge and gap for amorphous semiconductors: the <u>non-direct edge and non-direct gap</u>. For amorphous silicon hydride (a-Si:H), the optical absorption coefficient increased from less than 10 cm^{-1} to 10^4 cm^{-1} over 0.5 eV. The <u>low energy side of the non-direct edge</u> ($\alpha \leq 10^3$ cm^{-1}) exhibits an exponential or Urbach edge with a slope 5 to 20 times larger than the slope of the Urbach edge for either the direct or indirect edges. This large difference has been interpreted in the past as suggesting a different physical mechanism for the Urbach phenomena in amorphous semiconductors than in crystalline semiconductors and insulators. The <u>high energy side of the non-direct edge</u> ($\alpha \geq 10^3$ cm^{-1}) may be called the Tauc regime after the physicist who contributed so greatly to its understanding[4].

The commercial importance of films of a-Si:H in photovoltaics and electronics led to the development of deposition techniques that produce amorphous semiconducting films of "semiconductor grade" purity with extremely reproducible electrical and optical properties. The existence of such materials and comprehensive literature on their optical properties suggests theoretical opportunities. This paper focuses on some remarkable experimental connections between the optical properties of amorphous silicon hydride (a-Si:H) and crystalline silicon (c-Si) and is intended to further encourage the development of fundamental models for the optical absorption and optical gap of these technologically important semiconductors.

Embarking on a comparison of c-Si and a-Si:H immediately raises a question: How can one compare an alloy and an element?

This paper concentrates on optical data obtained from "solar-grade" films of a-Si:H produced by the glow discharge of pure silane. For such films, the hydrogen concentration is near 10%. It is very weakly dependent on growth conditions and is chiefly influenced by small changes in substrate temperature, which is in turn controlled in a narrow range about 250C. The optical properties of such films are not strongly dependent on small variations in hydrogen concentration and indeed it is accepted that the dominant effect of the small amounts of hydrogen in the lattice is through its effect on the structural disorder of the continuous random network[3]. We will also consider optical data on CVD films of amorphous silicon produced by thermal decomposition of silane at temperatures above 500C. These films have only trace amounts of hydrogen. Films of amorphous silicon hydride can be produced with considerably higher hydrogen concentrations by deposition from silane or disilane at lower substrate temperatures. They have optical properties that are strongly influenced by hydrogen bonding and fall outside the focus of this paper.

1.2 The Non-Direct Edge of Amorphous Semiconductors

The first step toward a quantitative theory of the optical absorption in amorphous semiconductors was the the "Independent Band Model". The model was introduced in 1966[5] and was summarized by Tauc[4] in 1972, and recently reviewed by Hass and Eherenreich[6] in 1985. In this model, the disordered, or amorphous, semiconductor is derived from the effect of site disorder on a postulated ordered semiconductor ("virtual crystal") with the same elemental composition and coordination. The valence (conduction) band wave functions for the amorphous semiconductor can be expanded in terms of the Bloch waves of the virtual crystal and can be derived from them through a unitary transformation. There is no mixing in energy of valence and conduction band wave functions in this model and the "forbidden gap" of the virtual crystal continues to separate the two bands of the amorphous semiconductor which despite arbitrary disorder never overlap. This assumption has most validity for small disorder, but paradoxically, the greatest success of the Independent Band Model has been in the area of the optical absorption edge of highly disordered amorphous semiconductors.

A significant experimental feature of the Independent Band Model is that it makes quantitative predictions. From this model it is possible to deduce the magnitude of the dipole matrix elements of both the amorphous semiconductor from the optical data if density of states (DOS) derived from experiment are available. The Independent Band Model leads a new sum rule, namely the invarience of the integral of the imaginary part of the dielectric constant over all optical transitions between the valence and conduction bands for the amorphous semiconductor _and_ its virtual crystal. Satisfaction of this sum rule is an important experimental confirmation of the critical assumption of the model, namely that the amorphous semiconductor is the disordered form of an ordered semiconductor or "virtual crystal".

The Independent Band Model is incomplete since it considers the effect of disorder on the optical absorption of the virtual crystal, but ignores the effect of disorder on the density of states of the virtual crystal. This inconsistency does not limit the experimental utility of the model. The experimentalist can choose an energy dependence of an empirical DOS for the amorphous semiconductor that fits the energy dependence to the absorption edge. The magnitude and energy dependence of the DOS can then be compared to that predicted by more fundamental theories of the effect of disorder on the band edges of ordered semiconductors.

The optical gap or ("Tauc gap") of the Independent Band Model is a fitting parameter that is a useful measure of the electrical forbidden gap of the amorphous semiconductor as well as its absorption edge. Unfortunately there has been no sustained theoretical effort to relate this gap to fundamental properties of the semiconductor, despite the wealth of experimental data on very well defined films of a-Si:H and more recently a-Ge:H. A fruitful research route might be to follow the approach of the Independent Band Model and examine the connection between the disordered semiconductor and its "virtual crystal".

This paper is an experimental exploration of that route. Optical data is presented and discussed that suggests that c-Si is an excellent candidate for the "virtual crystal" of a-Si:H. More precisely the paper explores the hypothesis that the "Tauc region" of the _non-direct edge_ of a-Si:H between 1.6eV and 3.2eV is derived from the wave functions in that part of the Brillouin Zone of c-Si that contributes to the <u>direct edge</u> near 3.2eV (the zone center $\Gamma_{25'}$ to Γ_{15} transition[7]). The hypothesis is in reasonable agreement with: (1) with the magnitude of the optical dipole matrix element of a-Si:H and c-Si at their absorption edges; (2) the magnitude and temperature dependence of the optical gaps for a-Si:H and c-Si; and (3) the magnitude and temperature dependence of the slope of the Urbach edge for c-Si and a-Si:H. These connections between the absorption spectra of c-Si and a-Si:H appear to have been overlooked and are strong enough to suggest theoretical opportunities. Further more they are consistent with recent fundamental theories of the effect of site disorder on the optical absorption of c-Si and other semiconductors with long range order.

It is interesting to note that more than 20 years ago, the paper[5] that first introduced the Independent Band Model speculated that the direct gap of c-Ge ($\Gamma_{2'}$ to $\Gamma_{25'}$ transition at 0.8eV) might be the "virtual crystal" for a-Ge. Recent optical data on a-Ge:H shows a Tauc gap of 1.07eV[8] and thus rules out the first direct transition at the Γ point as a candidate for the "virtual crystal" for a-Ge:H. An candidate might be the wave functions contributing to the peak in the imaginary part of the dielectric constant $\{\varepsilon_2(E)\}$ of c-Ge at 2.3eV corresponding to the "almost direct" transition at the Λ point[9]. Recent photovoltaic interest in multi-junction solar cells has generated comprehensive optical data on well defined a-Ge and a-Ge_xSi_{1-x} films[10] and would justify an exploration of that original hypothesis in this paper.

The silicon system is perhaps a better candidate for such speculations since its band structure is considerably simpler than the germanium system and it will be the focus of this paper although optical data on a-Ge will be introduced when appropriate to broaden the perspective. Despite this difference, the present paper can be considered a revisit of a speculation first advanced by Tauc, Grigorovici and Vancu[5] in 1966 for understanding the absorption edge of amorphous semiconductors in terms of the effect of disorder on their crystalline form.

1.3 Fundamental Models for the Absorption Edge

It is beyond the scope of this paper to discuss in any detail recent fundamental models for the optical properties of amorphous semiconductors. These theories start with a simplified Hamiltonian for the coupled phonon electron system. From an exact solution of the simplified Hamiltonian, they can, in principle, produce all the effect of thermal or static disorder on the absorption edge.

A particularly challenging task for such models is a first principles derivation of the Urbach Edge. As noted earlier in the introduction, amorphous and crystalline semiconductors, and high band gap insulators share a common feature in the low energy side of the absorption edge. In the range of absorption coefficients below 10^3 cm^{-1} (below the direct, indirect or non-direct "optical gap"), the optical absorption is exponential over three or more orders of magnitude. This region is called the Urbach edge region. In the thirty years since its discovery[11], the Urbach edge has intrigued theorists by its generality in such a wide range of materials[12]. A first principles calculation of the Urbach Edge would not only resolve the major theoretical question of the derivation of an exponential edge from gaussian disorder, but would also resolve the question of whether the Urbach edge in crystalline and amorphous semiconductors arise from the same physical mechanisms[12].

Recent theoretical work has been remarkably successful in modeling the Urbach edge phenomena in thermally or structurally disordered semiconductors and it is now generally accepted that either static or dynamic atomic scale disorder can produce an exponential character of the absorption edge largely through an exponential density of states (DOS) proceeding from the conduction and valence band tails[13]. Particularly relevant to this paper is the recent theory of Grein and John[14] which presents a first principles calculation of the Urbach edge. Their papers calculate the following experimental quantities for crystalline semiconductors and the disordered semiconductors derived from them: the magnitude and temperature dependence of the slope of the Urbach edge; the temperature dependence of the band gap; and the magnitude and temperature dependence of the coupling between the optical gap and the Urbach slope. As will be seen, this theory is remarkably successful in reproducing a wide range of experimental data on the absorption edge of a-Si:H and c-Si. The mathematics of the theory is complex, its physics is not. The success of the theory suggests that the optical absorption edge in amorphous and crystalline semiconductors can be understood in terms of the effect of thermal and structural site disorder on the band structure of an ordered semiconductor. Its success indirectly supports the contention of this paper that many of the features of the optical absorption edge of a-Si:H may be related to the effect of site disorder on the electronic states contributing to the direct gap of c-Si at 3.2eV (the zone center $\Gamma_{25'}$ to Γ_{15} transition).

2. QUANTITATIVE MODELS FOR THE ABSORPTION EDGE

2.1 Optical Absorption in Ordered Semiconductors

The optical properties of a material are determined by the magnitude and energy dependence of the complex dielectric constant, $\varepsilon(E) = \varepsilon_1(E) + i\varepsilon_2(E)$. Within the one electron approximation, it is easy to show that the imaginary part of the complex dielectric constant, $\varepsilon_2(E)$, is given by[3]:

$$\varepsilon_2(E) = (2\pi e)^2 (2/V) \sum |R_{v,c}|^2 \delta(E_v - E_c - E) \qquad (1)$$

where V is the illuminated volume of the sample, E is the energy of the incident light wave and $|R_{v,c}|^2$ the dipole matrix element. The summation is over the wave functions of the occupied (valence) band and unoccupied (conduction) <u>single spin band states</u> for the illuminated volume and T=0K is assumed for simplicity. The momentum matrix element $|P_{v,c}|^2$ can be used in an alternative formulation to Eq.(1), but experimental data suggest the dipole matrix element is a more appropriate choice for a-Si:H.

$R_{v,c}$ can be written for a semiconductor, with long range order, in a form that explicitly indicates conservation of crystal momentum, k(v,c):

$$R_{v,c} = (3)^{1/2} Q(E) \delta_{k(v),k'(c)} \qquad (2)$$

In Eq.(2) the factor of $(3)^{1/2}$ is introduced to compensate for a factor of $[1/3]$ to be introduced when an average is taken over polarization of the incident light. The quantity Q(E) has the same units as $R_{v,c}$ and exhibits its energy dependence.

From Eq.(1) and (2), $\varepsilon_2(E)$ for a ordered semiconductor can be written after averaging for non-polarized light:

$$\varepsilon_2(E) = (2\pi e)^2 (2/V) \sum Q^2(E) \delta(E_c(k)-E_v(k)-E) \qquad (3)$$

Integration of Eq.(3) over energy gives the dipole sum rule:

$$\left\{ \int \varepsilon_2(E)dE \right\}_x = (2\pi e)^2 <Q^2(E)> \nu \rho_a \qquad (4)$$

In Eq.(4), ν is the number of valence electrons per atom, ρ_a is the atomic density and $<Q^2(E)>$ is the average of $Q^2(E)$ over all optical transitions between the valence and conduction bands of the ordered semiconductor.

For parabolic bands in the vicinity of a direct gap, E_G^x, with $Q(E)=constant=Q_x$, the well known expression for the absorption edge of a direct gap crystalline semiconductor[1] can be readily derived from Eq.(3):

$$\varepsilon_2(E) = C^x (E - E_G^x)^{1/2} \qquad (5)$$

where $C^x = 1.37 Q_x^2 M_x^{3/2}$; M_x is the inverse arithmetic average of the density of states effective masses of the crystalline conduction and valence bands in units of electron mass, m_0; E_G^x is the gap separating the two bands in eV; and Q_x is the dipole matrix element in Å in the neighborhood of E_G^x. It is important to distinguish between Q_x defined by Eq. (5) and $\{<Q^2(E)>\}^{1/2}$ defined by Eq. (4). They are related but not equivalent quantities.

2.2 Optical Absorption for Disordered Semiconductors

In the Independent Band Model, the wave functions of the disordered semiconductor are formed from the wave functions of the ordered semiconductor by a unitary transformation. Applying such a transformation to Eq.(1) and <u>eliminating momentum conservation</u> in the transition leads easily to Eq.(6) which defines a first order optical transition in a semiconductor with no long range order[3,4,6]:

$$\varepsilon_2(E) = (2\pi e)^2 (2/VN_o) \sum Q^2(E) \delta (E_c - E_v - E) \qquad (6)$$

In Eq.(6), N_o is the total number of <u>single spin states</u> in the valence band for the illuminated volume, V, and the summation is over all conduction and valence band <u>single spin states.</u> The difference between the delta function in Eq. (6) and Kronnecker delta in Eq. (3) should be noted.

Replacing the summation in Eq.(6) by an integral leads to the well known expression for $\varepsilon_2(E)$ for amorphous semiconductors in terms of the "joint density of states" between the valence and conduction bands:

$$\varepsilon_2(E) = (2\pi e)^2 (2/VN_o) \int dZ\, Q^2(E)\, N_v(Z)\, N_c(Z + E) \qquad (7)$$

where $N_{c,v}(E)$ is the density of single spin conduction (valence) band states. The dipole matrix element, $Q(E)$, in Eq.(7) is the same function shown in Eq.(3) since matrix elements are invarient under the unitary transformation connecting the disordered semiconductor with its "virtual" crystal. Eq.(7) makes it possible to compute the magnitude of the optical absorption of a disordered semiconductor <u>if the "joint density of states" is known</u>.

Eq. (7) also leads to the identical dipole sum rule as in Eq.(4). Assuming that the electronic density ($v \rho_a$) of both the ordered and disordered semiconductor are the same, integration of Eq.(7) over energy, leads to:

$$\left\{ \int \varepsilon_2(E) dE \right\}_a = (2\pi e)^2 <Q^2(E)>\, v\, \rho_a = \left\{ \int \varepsilon_2(E) dE \right\}_x \qquad (8)$$

The critical question for the validity of the independent band model, as pointed out by Hass and Ehrenreich[6], is the equality of the integrals on the left and right sides of Eq.(8). As will be seen in the next section, the dipole sum integral for a-Si:H and c-Si are almost identical, suggesting that with respect to Eq.(8), c-Si is the "virtual crystal" for a-Si:H.

Eq.(7) can be evaluated for <u>parabolic bands</u> and constant dipole matrix element "Q_a" in the vicinity of the non-direct edge, $E_G{}^a$:

$$\varepsilon_2(E) = C^a (E-E_G{}^a)^2 \qquad (9)$$

where $C^a = Q_a{}^2 M_a{}^3 (0.2/v)$; v is the number of valence electrons per atom; M_a is the <u>geometric mean</u> of the density of states effective mass in units of free electron mass; $E_G{}^a$ is the energy gap in eV separating the conduction and valence bands of the disordered semiconductor; and Q_a is the dipole matrix element in Å. Eq.(9) was first presented by Tauc et al.[5], in a form appropriate for a <u>constant momentum matrix element</u>. The gap $E_G{}^a$ is often denoted the "Tauc gap" and the region over which Eq.(9) applies is denoted the Tauc regime of the optical absorption coefficient. Eq. (9) is the equivalent of Eq.(5) for a <u>first order non-direct optical transition</u>.

The next section presents a variety of experimental data that test the hypothesis that the absorption edge of a-Si:H can be derived from the effect of disorder on the direct edge of c-Si.

3.0 OPTICAL ABSORPTION IN c-SI AND a-SI:H

3.1 Dipole Sum Rule

Fig.(1) is a plot of the imaginary part of the dielectric constant for low pressure CVD a-Si: and c-Si over a 6 eV range[15]. The areas under the two curves are equal to 62eV from 0-6eV. Making the reasonable assumption that there is very little difference between a-Si: and c-Si up to 20eV, Eq. (8) leads to $\{<Q^2(E)>\}^{1/2} = 1$Å as an average over all the valence electrons. This magnitude is in excellent agreement with estimates made from the oscillator strength sum rule for an oscillator at 4.2eV i.e. the broad maximum in $\varepsilon_2(E)$ for c-Si. As pointed out by Hass and Ehrenreich[6], $\{<Q^2(E)>\}^{1/2} = 1$Å is also in agreement with first principles calculations of dipole moments for silicon. It is important to note that $\{<Q^2(E)>\}^{1/2}$ is an average dipole matrix element over all optical transitions from the valence band and is not equal to Q_a (Eq.(9)) or Q_x(Eq.(5)) which are constant matrix elements defined over a limited energy range. However, it is anticipated that all

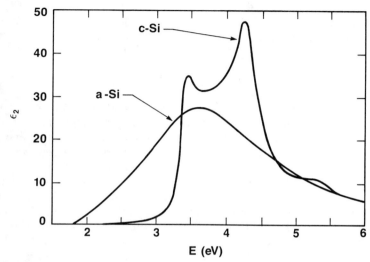

Figure 1. Imaginary part of complex dielectric constant for a-Si:H and c-Si (adapted from reference (3))

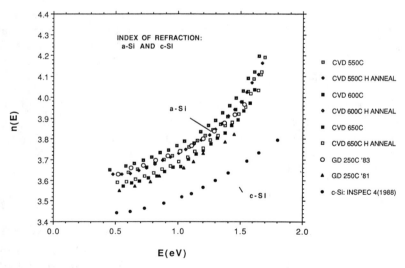

Figure 2. Magnitude and dispersion of index of refraction of CVD and Glow Discharge a-Si:H compared to that of c-Si[19]

these quantities will of the order of atomic dimension within the solid; i.e ≤ 2.4Å. Hass and Ehrenreich[6] have shown that Eq. (8) is satisfied for a number of crystalline/amorphous semiconductor pairs that have the same symetry and where there is sufficient optical data for the comparison.

3.2 Dispersion of the Index of Refraction of c-Si And a-Si:H

The global similarity between $\varepsilon_2(E)$ for c-Si and a-Si:H shown in Fig. 1 has consequences for the magnitude and dispersion of the index of refraction for a-Si:H and c-Si. Figure 2 exhibits the energy dependent index of refraction for eight a-Si:H films prepared by two deposition techniques: chemical vapor deposition (CVD) with or without H anneal at substrate temperatures between 550C and 650C and glow discharge (GD) amorphous silicon films of solar cell quality prepared at a substrate temperature of 250C in 1981 and 1983. As shown in Table 1, the films exhibit Tauc gaps that range from 1.4 to 1.7eV, spin densities that range from 10^{19}/cc to 10^{15}/cc, and magnitudes for E_O (the inverse slope the Urbach edge) that range from 40meV to 120meV (in this case only the glow discharge films could be described as "semiconductor grade"!).

Despite this wide variety of behavior in absorption, the index of refraction data for the a-Si:H films shown in Fig.2 are remarkably similar. The magnitude and dispersion of the index of refraction for c-Si is shown in Fig. 2 over the same energy range and is within five to ten percent of that of a-Si:H from 0.5eV to 1.7eV.

The reason for this agreement is simple to understand and model from the data for $\varepsilon_2(E)$ shown in Fig.1. Approximating the peak in the experimental data by a <u>delta function resonance absorption</u> at an energy, E_R, it is easy to show from the Kramers-Kronig equations and the plasma sum rule that[3]:

$$n^2(E) = \varepsilon_1(E) = 1 + \{E_P^2/\{E_R^2 - E^2\}\} \tag{10}$$

where E_P, is the plasma energy $E_P = (h^2 \nu \rho_a e^2/\pi m_o)^{1/2} = 16.6eV$ for c-Si, and the other quantities have been previously defined Wemple and DiDomenici introduced an additional degree of freedom in Eq. (10) by replacing E_P^2 by $E_D E_R$ where E_D is a quantity that is proportional to the density. Clearly from Fig. 1, "E_R" should be approximately the same for c-Si and a-Si:H with a magnitude of ≈4.0eV.

Fig.3 displays the data of Fig.2 as a linear plot of $1/\{n^2(E)-1\}$ as a function of E^2 as suggested by Eq.(10). The agreement with Eq.(10) is excellent and the figure highlights the similarity between c-Si and a-Si:H with respect to the index of refraction for a a-Si:H films with a broad range of properties. Table 1 displays the constants E_R and E_P. The differences between c-Si and a-Si:H are very small and can largely be accounted for by the shift in $\varepsilon_2(E)$ for a-Si:H to lower energies seen in Fig. (1). It is interesting to note that the highest quality film of a-Si:H in the table as measured by solar cell performance (GD 250C 83) has a value for E_P only 4% less than that for c-Si suggesting equivalent electron densities. It is not surprising that E_P is only 80% of the plasma energy of c-Si given the crudeness of the approximation leading to Eq.(10)

The index of refraction of a-Si:H is dominated through the K-K relation by the strong absorption at 3.4eV and the index of c-Si by the broad peak between 3.5 and 4.5eV. The magnitude and dispersion of the index of refraction is thus little more than a reflection of the peak structure in $\varepsilon_2(E)$. Any significant difference between experimental data and the average constants shown in Table 1 can usually be traced to significant departures in density from ideal values due to porosity of the films or artifacts of the measurement due to poor surface quality of the films. The effects of temperature and pressure on the index and its dispersion for both c-Si and a-Si:H should be related to the shift in the peak of $\varepsilon_2(E)$ with temperature or pressure ("E_R") or from the direct effects of temperature or pressure on the density, through the quantity "E_P^2".

3.3 Absorption at the Non-Direct Edge of a-Si:H

Eq.(9) is the fundamental result of the independent band model specialized to parabolic bands and a constant dipole matrix element and should apply in the vicinity of the band edge. In this section we use Eq.(9) to deduce optical parameters of a-Si:H in the region of the non-direct gap for subsequent comparison with optical parameters of c-Si in the region of its direct gap (section 3.4).

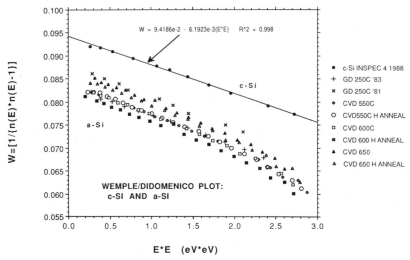

Figure 3. Index of refraction data from Figure 2 plotted to exhibit functional dependence of Equation (10) of text.

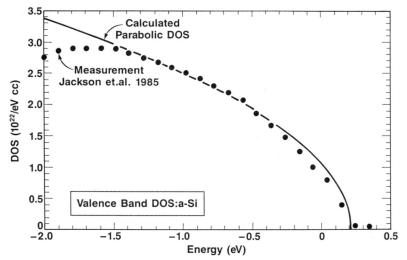

Figure 4. Experimental valence band density of states replotted from data of Jackson et al[16] compared to a parabolic energy dependence

How reasonable is the assumption of parabolic bands used in deriving Eq.(9)? Fig. 4 and 5 display the direct measurements by Jackson et al[16] of the single spin density of states of the valence and conduction bands of a-Si:H over a comparable energy range. The data <u>has been fitted</u> to a parabolic density of states. The fitting parameters are: a density of states effect mass of 4.06 m_e for the conduction band, 3.54 m_e for the valence band, and an energy gap of 1.66eV. Thus from the experimental data from 1.7 to 4.0 eV the joint density of states is well approximated by the convolution of two parabolic bands as given in Eq.(9), and from the fit $M_a=3.8m_e$.

How good is the approximation of a constant matrix element in the derivation of Eq.(9)? Fig. 6 displays $(\varepsilon_2(E))^{1/2}$ as a function of photon energy for a-Si:H[17]. The data was extracted by transmission measurements on a-Si:H films of different thicknesses prepared by glow discharge decomposition of silane on substrates held at 250C. The data is in excellent agreement with Eq.(9) over the range 1.6 to 3.2eV and indicates an energy gap of 1.64eV and a magnitude for the coefficient C_a of 9.36 eV^{-2}.

Figs. 4, 5 and 6 support the assumption of a constant matrix element and a parabolic density of states leading to Eq.(9) and justify using this expression to obtain the optical parameters of a-Si:H. The data in these figures is consistent with a value for the dipole matrix element of $Q_a=1.9Å$; an average effective mass of $3.8m_o$ between 1.6eV and 3.2eV; and a Tauc gap of 1.64-1.66eV for "semiconductor grade" a-Si:H.

The magnitude of the dipole matrix element derived from the fit is of the order of atomic dimensions as anticipated. At first glance an effective mass almost 4 m_e for amorphous silicon is disturbing; an effective mass of the order of 1 m_e is usually assumed in the interpretation of tunneling and transport measurements for amorphous superlattices based on a-Si:H[18]. However it is should be remembered that the mass parameter in Eq.(9) is not a <u>transport mass</u>, but rather a <u>density of states mass</u> that reflects the number of single spin states per unit volume available in the energy shell corresponding to the incident photon. It is not a dynamic or inertial quantity. Indeed it well be seen in the next section that density of state masses of the order of four or

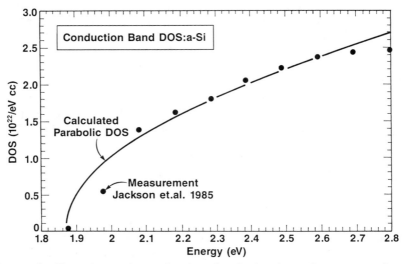

Figure 5. Experimental conduction band density of states replotted from data of Jackson et al[16] compared to a parabolic energy dependence

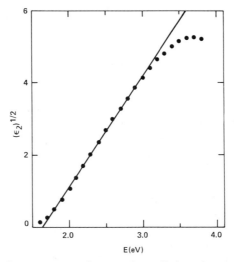

Figure 6. Imaginary part of complex dielectric constant of a-Si:H plotted to exhibit correspondence with Equation (9) of text.

more can be associated with the direct gap of crystalline silicon at 3.2eV.

The observation that the density of states follows an approximate parabolic law over such a wide energy range is puzzling and has yet to receive a fundamental explanation. It is tempting to associate the parabolic density of states with the immediate effect of disorder on a direct gap before the transition to the exponential Urbach edge. Given the universality of the Urbach edge, the parabolic dependence of the density of states <u>in the transition to the exponential regime</u> might be equally universal.

3.4 Absorption at the Direct Gap of c-Si

Figure 7 is a plot[15] of the absorption coefficient for crystalline and amorphous silicon at 300K over the same range of energies as in Fig. 1 (1.0eV to 6.0eV). The indirect gap of crystalline silicon silicon is indicated at 1.11eV, as is the non-direct (Tauc gap) gap of amorphous silicon at 1.64eV. Neither of these two gaps are prominent "features" in the absorption coefficient. For a-Si:H the only significant feature is the transition from "band-to-band" absorption to "dangling bond-to-band" absorption near 1.4eV. For c-Si the most significant feature of the absorption curve is the sharp rise near 3.4eV, the direct gap at the zone center $\Gamma_{25'}$ to Γ_{15} optical transition.

Fig. 8 exhibits an expanded plot of $\varepsilon_2(E)$ ($\equiv 2n(E)k(E)$) in the region of the direct gap for c-Si[19]. The solid curve in Fig.8 is a fit of Eq.(5) to the data over the range 3.2eV to 3.4eV and $C_x = 80.9$ ($\equiv Q_x^2 m_x^{3/2}$). The hypothesis that the non-direct edge of a-Si:H is derived from the direct edge of c-Si suggests that $Q_x = Q_a = 1.9$Å and that $M = M_a = 3.8 m_0$. Substitution in the expression for C_x leads to $C_x = 26.7$ in approximate agreement with the experimental value of 80.9.

Analysis of the magnitude and energy dependence of the non-direct edge of a-Si:H {Fig.6 and Eq. (9)} and the direct edge of c-Si {Fig.8 and Eq. (5)} suggests that the density of states effective mass and the dipole matrix element for the two transitions are of the same magnitude. It is supportive of the hypothesis that the non-direct edge is the disorder broadened direct edge as suggested

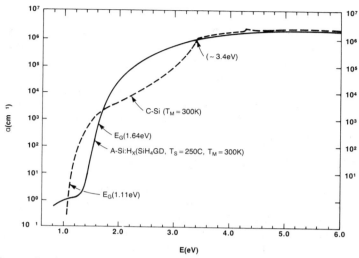

Figure 7. Optical absorption coefficient of c-Si[19] and a-Si:H[15].

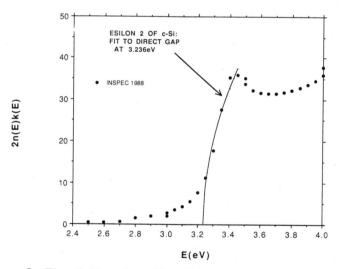

Figure 8. Fit of Equation (5) of text to the imaginary part of the complex dielectric constant, $\varepsilon_2(E)$ {=$2n(E)k(E)$}, at the direct gap of c-Si[19] near 3.2eV.

by Fig. 7. The next section compares the Urbach edge of c-Si and a-Si:H with respect to this same hypothesis.

3.5 Urbach Edge of c-Si and a-Si:H

Perhaps the most remarkable similarity between crystalline and amorphous silicon is in the Urbach edge of the two materials. Fig.9 presents an expanded plot of the absorption coefficient of c-Si between 1.01eV and 1.21eV[20]. The data is in excellent agreement with earlier data of Mac Farlane et al.[21] The changes of slope in the data exhibited in Fig. 9, marked with arrows at 1.172eV and 1.057eV can be ascribed to the excitation of a valence electron to an exciton 14.7meV below the conduction band with the emission or absorption of a transverse optical phonon of 58meV[22]. These transitions have been directly observed in modulation spectroscopy[23]. The lower feature at 1.03eV, seen in Fig.9, has not been pointed out previously. It is either another phonon assisted transition or an experimental artifact related to the extremely low levels of absorption below 1.03eV for the 970μ crystal. Fig.10 exhibits the absorption coefficient in the Urbach region between 1.01eV and 1.07eV. Exponential behavior over almost three orders of magnitude is apparent.

In the experimental literature on the Urbach edge it is customary to express the absorption coefficient as:

$$\alpha = \alpha_o \exp(E-E_F)/E_o$$

where E_F is the Urbach focus and E_o is the Urbach edge parameter. From Fig. 10, the Urbach parameter, E_o, of c-Si is 10 meV at T=300K. This result can be compared with the recent theoretical model of Grein and John[14]. Their model ascribes the Urbach edge to the effect of thermal disorder on the electronic states which induces an exponential density of states in the band tail of a free carrier band. From an exact solution to an approximate Hamiltonian which couples phonons contributing to the disorder to the electronic energy levels, they predict, for c-Si, an Urbach slope of $E_O=8.6$ meV. This result is in excellent agreement with experiment given that the only free parameter of the theory is the valence band deformation potential, E_d which is set equal to its Shockley-Bardeen[27] value for Si ($E_d=11.3eV$).

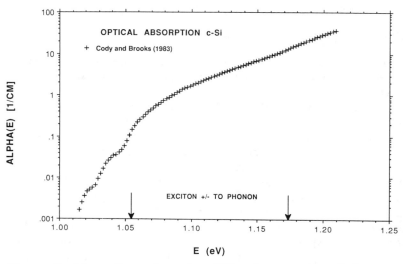

Figure 9. Absorption edge of a 970 micron wafer of P-type silicon (111) measured in transmission[20].

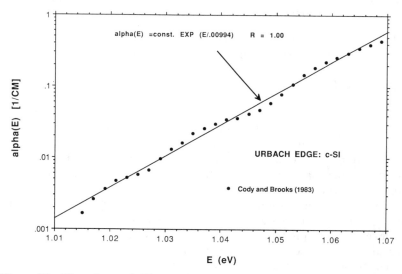

Figure10. The data of Figure 9. plotted on an expanded energy scale (1.015eV-1.07eV) to exhibit Urbach edge below the indirect gap.

The Urbach edge in c-Si arises from thermal disorder alone; there is no significant static disorder. For a-Si:H, however, there is both thermal and static disorder. Fig.11 exhibits the Urbach edge for two films of a-Si:H. For a-Si:H, E_0=50meV, about five times that of c-Si! The large difference suggests a different physical mechanism as does the fact that many amorphous glasses exhibit a similar magnitude for E_0. However it has been shown experimentally for a-Si:H that <u>static and thermal disorder make additive</u> contributions to E_0[24,4]. This conclusion is supported by the Grein-John[14] theory. In general: $E_0 = E_0^T + E_0^x$ where E_0^T is the contribution to the Urbach edge from thermal disorder and E_0^x is the contribution from static disorder which is temperature independent. Thus the relative constancy of E_0 for glasses may reflect a similar state of structural disorder produced by similar temperatures of synthesis.

For c-Si: $E_0=E_0^T=10$meV. For a-Si:H: $E_0=45$meV. Measurements of the effect of temperature on the absorption coefficient for films identical to those shown in Fig. 11 indicate that for a-Si:H: $E_0^T = 17$meV at 300K[25] or about 1.7 times larger than the experimental value of for c-Si of $E_0^T=10$meV. This difference is easily accounted for by a relatively small (30%) increase in the deformation potential of a-Si:H compared to c-Si since from the Grein-John theory[14]: $E_0^T \approx (E_d)^2$.

Finally it is worth noting that the magnitude of the derivative $(\partial E_G/\partial E_O)|_{T=300K} = G$, is a constant=7.2 in the theory of Grein and John[14] for both c-Si and a-Si:H for changes <u>in either thermal or static disorder</u>. Indeed in this theory, G is a numerical constant for semiconductors at room temperature with the similar properties as c-Si. $\{G=(\pi/(2)^{1/2}0.307)=7.2\}$ The magnitude of G is in excellent agreement with the experimental value of 6.2 obtained for a-Si:H[3] and a-Ge:H[8] against changes in <u>both structural and thermal disorder</u>. This result is a remarkable confirmation of the experimental observation that structural and thermal disorder enter on the same footing for both the Urbach parameter E_0 and the Tauc gap E_G.

In the next section we consider the effect of temperature on the in-direct and direct edges of c-Si and the non-direct edge of a-Si:H as another point of comparison between the two materials.

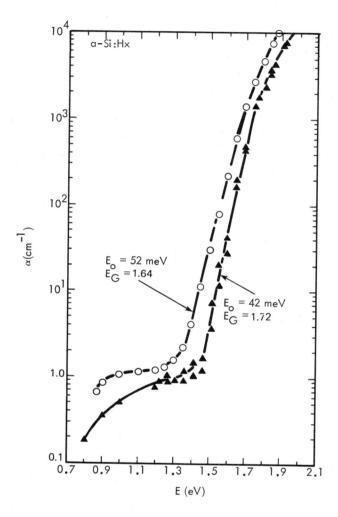

Figure11. Exponential or Urbach edge of two films of a-Si:H (GD 250C 81 and 250C 83 of Table 1). E_G is the Tauc gap defined by Eq.(9) and E_o is the Urbach edge parameter defined in Section 3.5 of the text.

3.6 The Effect of Thermal Disorder on the Band Edge

The goal of this section is to introduce a simple physical model for the effect of temperature on the direct, indirect and non-direct band gaps of a semiconductor. In the next section, this model will be used to compare $(\partial E_G/\partial T)$ for a-Si:H and c-Si. Despite its simplicity, the model captures the basic physics of phenomena sufficiently well enough to be an excellent <u>one parameter fit</u> to experimental data on the temperature dependence of the band edge for a wide range of crystalline and amorphous semiconductors[3].

The starting point for the model is a Taylor series expansion for the band gap at constant volume. Under this condition

$$\Delta E_G(T) = E_G(0) - E_G(T) = (D/2)\{<u^2>_T - <u^2>_0\} \qquad (11)$$

where the quantity "u" $=R-<R>$, "R" is a generalized coordinate coupled to the band gap through a net <u>second order</u> deformation potential "D", and $<u>=0$ from the assumption of constant volume. As pointed out by Allan and Cardona[26], the quantity "D" contains terms that are proportional to the square of the usual deformation potential introduced by Schockley and Bardeen[27], E_d as well as intrinsic second order terms. In what follows, it will be assumed that $D \approx E_d^2$ since from Grein and John[14], this term is sufficient to account for the effect of thermal disorder.

From Eq.(11) the shift in the band gap at <u>constant volume</u> is proportional to the thermal energy of the semiconductor. There is another contribution to the thermal shift of the band edge: thermal expansion. Through the Gruneisen model for thermal expansion this contribution to the temperature shift of the band edge is again proportional to the <u>total thermal energy of the material</u>. An Einstein model for the thermal energy in the solid is utilized in the derivation since it includes the same physics as the more realistic Debye model but is mathematically more convenient. Combining the constant volume term, Eq.(11) and the effect of thermal expansion, a universal expression for the effect of temperature on the optical gap of semiconductors is obtained[3]:

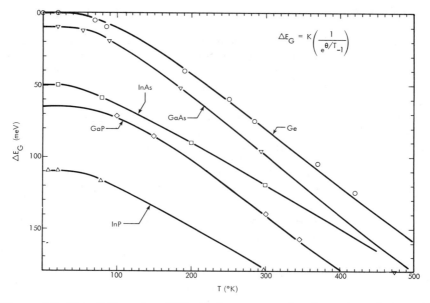

Figure 12. Fit of Equation (12) of the text to the thermal variation of the band gap for elemental and compound crystalline semiconductors. See Table 2 for the source of the data.

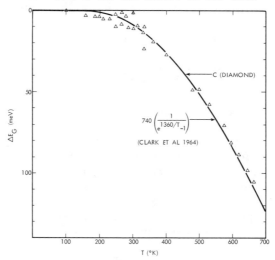

Figure 13. Fit of Equation (12) of the text to the temperature variation of the indirect optical gap of carbon[28] (diamond).

$$\Delta E_G(T) = K/\{\exp(\theta_E/T)-1\} \qquad (12)$$

where θ_E is the Einstein temperature.

The quantity $K = K_V + K_P$, where K_V is proportional to the the quantity D (or E_d^2) and is intrinsically positive. K_P includes the effect of thermal expansion on the band gap and has the same sign as that of the pressure coefficient of the band gap. With a Gruneisen model for thermal expansion it is easy to show that:

$$K_P = (\partial E_G/\partial P)_T \, (3R\theta_E \, \Gamma/V_A) \qquad (13)$$

In Eq.(13), $(\partial E_G/\partial P)_T$ is the isothermal pressure derivative of the optical gap in question, R is the gas constant, Γ is the Gruneisen constant, and V_A is the gram atomic volume. In general $K_V >> K_P$.

Fig. 12 and Fig 13 shows the fit of Eq.(12) to the temperature variation of the optical gap for a variety of crystalline semiconductors with either direct or indirect gaps[7,28]. The fit of Eq.(12) to the data is excellent and considerably better than that due to Varshni[29]. Table 2 summarizes the fitting constants and compares them with theory. It is apparent from the data that an excellent assumption for θ_E is 0.6 θ_D, where θ_D is the Debye temperature of the semiconductor which makes Eq.(12) a one parameter fit. This numerical relation is close to that expected on the basis of zero point energy considerations ($\theta_E = 0.75\theta_D$). From the table we note that constant volume effects (K_V) dominate the temperature derivative of the optical gap ($K_V >> K_P$).

There is nothing in the above model that restricts Eq.(12) to crystalline or ordered semiconductors and in the next section it will be applied to amorphous semiconductors such as a-Si:H and a-Ge.

3.7 Temperature Dependence of the Absorption Edge of a-Si:H and c-Si

Fig.14 is a plot of the temperature dependence of the indirect gap of crystalline silicon as determined by Bludeau et al[22] from wave length modulation. The full curve is Eq.(12) with K=0.1eV. Fig 15 is a plot of the indirect gap of amorphous silicon with

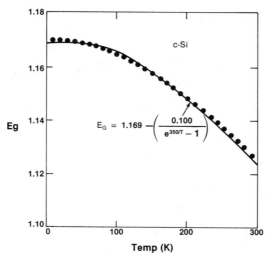

Figure 14. Fit of Equation (12) of the text to the temperature variation of the indirect band gap of c-Si as measured by Bludeau et al[22]

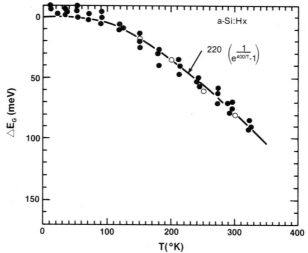

Figure 15. Temperature variation of the Tauc gap of a-Si:H (Equation (9)) fitted to Equation (12).

K=0.2eV. The difference of a factor of two in the quantity, K, between c-Si and a-Si:H can not be accounted for by differences in thermal expansion (compare {K_P/K_V} in Table 2). Assuming that $K_V \approx E_d^2$, the larger temperature derivative of the Tauc gap of a-Si:H compared to the temperature derivative of the indirect gap of c-Si requires a 40% larger deformation potential for a-Si:H. It is not accidental that this is of the same order as that necessary to account for the difference in thermal contributions to E_o between a-Si:H and c-Si in the model of Grein and John[14].

Grein and John[14] have derived the temperature dependence of the band edge of c-Si and a-Si:H from a first principles model. The key physical parameters of the theory are: (1) the dimensionless electron-acoustic phonon coupling constant, $S_{ac} \equiv \{E_d^2/2Mu^2 \, k\theta_D\}$, where E_d is the deformation potential, M is the atomic mass of the unit cell, k is Boltzman's constant, u is the velocity of sound and θ_D is the Debye temperature; and (2) the non-adiabaticity parameter, $\gamma \equiv \{2mu^2/k\theta_D\}$, where m is the dynamic (not density of states!) effective mass of the electrons.

A connection can be easily made between this fundamental theory and the simple physical model summarized in Eq. (12). For values of the constants comparable to those for crystalline silicon it can be shown that the constant volume parameter of Eq. (12), K_V, is given by:

$$K_V \cong (\pi/\sqrt{2}) S_{ac} \, \gamma k\theta_E \equiv 2.2(m/M) \, (E_d^2/k\theta_D)(\theta_E/\theta_D) \qquad (14)$$

Substituting an effective mass of 1 m_e (dynamic mass!), a unit cell mass of 56 atomic masses, a Debye temperature of 645K, and the Bardeen Shockley[27] valence band deformation potential of 11.3eV, $K_V \approx 0.04$ in reasonable agreement with the data for c-Si shown in Table 2 ($K_V \approx 0.1$) given the crudeness of Eq.(12). As will be seen, the direct calculation of Grein and John gives better agreement.

Fig.16 exhibits the predicition of Grein and John[14] for the temperature dependence of the indirect gap of c-Si ($E_d=11.3eV$, $\Theta_D=645K$, $M=56m_o$) compared to the experimental data of Bludeau et al[22]. Fig.17 compares the Grein-John[14] calculation of the temperature dependence of the non-direct gap of a-S:H ($E_d^a/E_d^x \cong 1.5$) with the data[3] shown in Fig. 15. It is noteworthy that

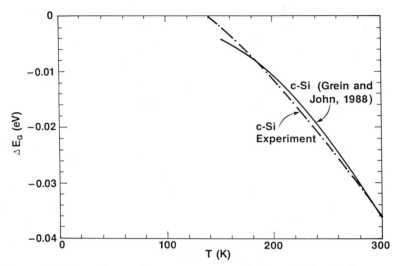

Figure 16. Experimental temperature variation of the indirect gap of c-Si after Bludeau et al[22] compared to the theoretical calculation of Grein and John[14].

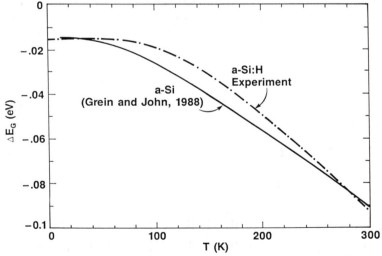

Figure 17. Experimental[3] of the non-direct (Tauc gap) of a-Si:H compared to the theoretical calculation of Grein and John[14].

this very simple model when solved accurately, with highly controlled approximations, can quantitatively account for the temperature dependence of the absorption edges of such complex materials as c-Si and a-Si:H. It would be of interest to see such calculations extended to c-Ge and a-Ge:H as well as c-Si$_x$Ge$_{1-x}$.

The persistent 40% increase in the deformation potential of a-Si:H compared to c-Si may arise from several causes. The increase in the deformation potential for the disordered semiconductor might be due to bond angle distortions arising from strains within the random network. It must be noted that the Grein-John[14] theory has been applied to the Urbach edge developed from the indirect gap, and the temperature dependence exhibited in Fig.16 is that of the indirect edge. An alternative explanation for $E_d^a/E_d^x \cong 1.5$ thus might be that the wave functions from which the non-direct edge of a-Si:H are derived have a 40% larger deformation potential than the deformation potential associated with the wave functions associated with the indirect edge of c-Si.

This last explanation is easily dealt with. Assuming that the deformation potential for a-Si:H is identical to that for the direct gap of c-Si ($\Gamma_{25'}$ to Γ_{15} transition at 3.2eV), a deformation potential ratio of 2 would be expected from the ratio of the pressure coefficients for the direct and indirect gap of c-Si[7]. This is more than enough to account for $K^a \approx 2K^x$. However, this explanation predicts that the temperature dependence of the $\Gamma_{25'}$ to Γ_{15} transition would be of the same magnitude as that of the non-direct edge of a-Si:H. Unfortunately for the explanation, there is no evidence that the thermal shift in the direct gap of c-Si is any larger than that for the indirect gap[30].

The "strained bond" explanation for $E_d^a/E_d^x \cong 1.5$ is also unsupported by any additional evidence. There is ample evidence that a-Ge has a considerably more strained random network[31] than a-Si; if strain were playing a role in enhancing E_d^a for a-Si, an even larger enhancement for a-Ge would be expected. A comparison of $(\partial E_G/\partial T)_V$ for c-Ge and a-Ge indicates a 30% larger thermal shift in the non-direct absorption edge of a-Ge compared to that for the indirect edge of c-Ge (Table 2). If thermal expansion were again ignored ($K_V \gg K_P$), this would imply $(E_d^a/E_d^x) \approx 1.2$ for Ge. If strain has such a small effect in a-Ge, an even smaller effect would be expected in a-Si:H.

It is clear from the last three sections that the temperature dependence of the absorption edge in amorphous and crystalline materials deserves further experimental and theoretical attention. A particularly interesting question is what might be expected for the variation of $(\partial E_G/\partial T)_V$ in amorphous and crystalline GeSi alloys, or amorphous layered Ge/Si films

4.0 SUMMARY AND CONCLUSION

This paper has focussed on simple physical models for the optical edge that were developed in the late seventies and early eighties. For some time it was thought that the success of these simple models in fitting experimental data was accidental given the complexity of the underlying phsics. The remarkable agreement between recent theories of the effect of thermal and structural disorder on the absorption edge and experiment summarized in this paper supports the insight of the earlier work that the optical properties of amorphous semiconductors can be derived from the effect of disorder on crystalline semiconductors. It encourages the attempt to find fundamental, first principles, models that address: the magnitude of the Tauc gap; the effect of pressure on the Urbach edge; the magnitude of the absorption coefficient near the edge including the energy dependence and magnitude of the dipole matrix element; the functional dependence of the absorption constant in the regime above the exponential edge, in particular, what statistical features produce the remarkable parabolic density of states exhibited in Fig. 2 and 3; and finally, the magnitude of the Urbach focus and its theoretical significance.

Finally the significant increase in the deformation potential for a-Si:H compared to c-Si, and for a-Ge:H compared to c-Ge, suggests further experimentation on that paradoxical phenomena: the effect of thermal disorder on highly disordered semiconductors. In addition to resolving theoretical uncertainties, such research might also contribute to thin film materials with improved device performance. The experimental observation from the Grein John theory that a-Si:H has a significantly larger valence band deformation potential than c-Si has important device implications.

ACKNOWLEDGEMENTS

The author is grateful: to present and past colleagues at Exxon's Corporate Research Laboratories for rewarding experimental collaborations in the study of amorphous silicon films: Bonnie Brooks, Ben Abeles, Chris Wronski, Tom Tiedje, Al Ruppert, and Peter Persans; to Bernie Silbernagel and Layce. Gebhard for determining the spin densities of the CVD films and Bernie Seraphim and Peter Hay of the University of Arizona for their preparation; to Henry Ehrenreich for numerous illuminating discussions on the fundamental features of the independent band model; to Chris Grein and Sajeev John of Princeton University for supplying preprints of their work prior to publication and for helpful discussions on the physics of their model for the Urbach edge. Finally the critical reading of this manuscript by Ben Abeles was extremely helpful in clarifying the ideas presented in this paper.

---

[1] Pankove, J., Optical Processes in Semiconductors, (Prentice-Hall, Englewood Cliffs, New Jersey, 1971).

[2] Wooten, F., Optical Properties of Solids, (Academic Press, New York, 1972).

[3] Cody, G.D., "The Optical Absorption Edge of a-Si:H" in Hydrogenated Amorphous Silicon, Semiconductors and Semimetals, 21C, ed. J. Pankove, (Academic Press, New York, 1984) p.11

[4] Tauc, J., "Optical Properties of Non-Crystalline Solids" in Optical Properties of Solids, ed. F. Abeles (North Holland, Amsterdam, 1972) p.279; Tauc J., "Optical Properties of Amorphous Semiconductors" in Amorphous and Liquid Semiconductors, ed. by J. Tauc (Plenum, New York, 1974) p. 159.

[5] Tauc, J. Grigorovici, R. and Vancu, A., Phys. Status Solidi **15**, 627 (1967).

[6] Hass, K. C. and Ehrenreich, H., Annals of Physics **164**, 77 (1985)

[7]Landoldt-Bornstien: New Series, **17**, "Semiconductors" ed. O. Madelung (Springer, Berlin, New York, 1982).

[8]Persans, P. D., Ruppert, A. F., Chan, S. S., and Cody, G. D., Solid State Communications, **51**, 203 (1984).

[9]Cardona, M., <u>Modulation Spectroscopy</u> (Academic Press, New York, 1969) p.244.

[10]Aljishi, S., Smith, Z. E., and Wagner, S.,"Optoelectronic Properties and the Gap State Distribution in A-Si, Ge Alloys" in <u>Amorphous Silicon and Related Materials</u> **B** ed. by H. Fritzsche, (World Scientific, Singapore, New Jersey, London and Hong Kong, 1989), p.887.

[11]Urbach, F., Phys. Rev. **92**, 1324 (1953).

[12]Hopfield, J. J., Comments Solid State Physics, **1**, 16 (1968); Kurik, M. V., Phys. status Solidi A, **8**, 9 (1971); Sa-Yakanit, V. and Glyde, C. H., Comments Solid State Physics, **13**, 35 (1987).

[13]John, S., Soukoulis, C., Cohen, M. H., Economou, E.N., Phys. Rev. Lett. **57**, 1777 (1986); John, S., Chou, M.Y., Cohen, M. H., and Soukoulis, C. M., Phys. Rev. B **37**, 6963 (1988); Sa-Yakanit, V., Phys. Rev. B **19**, 2266 (1979); Sritrakool, W., Sa-Yakanit, V., and Glyde, H. R., Phys. Rev. B **33**, 1199 (1986).

[14]Grein, C. H. and John, S., Phys. Rev. B **35**, 7457 (1987); Grein, C. H. and John, S., Phys. Rev. B **39**, 1140 (1989); Grein, C. H. and John, S., to appear in Solid State Communications.

[15]Adapted from Cody, G. D., in <u>Physics of Disordered Materials</u>, ed. D. Adler, H. Fritzsche and S. Ovshinsky (Plenum, New York, 1985) p. 327.

[16]Jackson, W. B., Kelso, S. M., Tsai, C. C., Allen, J. W., and Oh, S.-J., Phys. Rev. B **31**, 5187 (1985).

[17]Cody, G. D., Brooks, B. G., and Abeles, B., Solar Energy Mat. **4**, 231 (1982).

[18]Abeles, B., "Amorphous Semiconductor Superlattices" to be published in Proceedings of the Int. Conf. on Superlattices, Microstructures, and Microdevices (Trieste August 8-12, 1988).

[19]Taken from Aspnes, D.E., "Optical Functions (Complex Refractive Index and Absorption Coefficient" Properties of Silicon Inspec Data Review Series No.4 (INSPEC, London and New York, 1988).

[20]Cody, G. D. and Brooks, B.G., unpublished (1983); in tabular form in reference 19; see also Tiedje, T., Yablonovitch, E., Cody, G.D. and Brooks, B. G., IEEE Trans. on Elect. Devices **ED-31**, 711 (1984).

[21]MacFarlane, G. E., McLean, T. P., Quarrington, J. E. and Roberts, V., Phys. Rev. **111**, 1245 (1958).

[22]Bludeau, W., Onton, A., and Heinke, H., J., Appl. Phys. **45**, 1846 (1974).

[23]Frova, A., Handler, P., Germano, F. A. and Aspnes, D. E., Phys. Rev. **145**, 575 (1966).

[24]Cody, G. D., Tiedje, T., Abeles, B., Brooks, B. and Goldstein, Y., Phys. Rev. Let **47**, 1480 (1981).

[25]Tiedje, T. and Cebulka, J. M., Phys. Rev.B **28**, 7075 (1983).

[26]Allen, P.B. and Cardona, M., Phys. Rev. B **23**, 1495 (1974).

[27]Bardeen, J. and Schockley, W., Phys. Rev. **80**, 72 (1950).

[28]Clark, C.D., Dean, P.J. and Harris, P.V., Proc. Roy. Soc. A **277**, 312 (1964)

[29]Varshni, Y. P., Physica **39**, 149 (1967).

[30]Zucca, R. and Shen, Y., Phys. Rev.B **1**, 2668 (1970)

[31]Persans, P. and Ruppert, A. F. J. Appl. Phys. **59**, 271 (1986).

Table 1: Optical parameters for a-Si:H and c-Si

MATERIAL	GROWTH CONDITION	$E_G^{(1)}$ (eV)	$E_R^{(2)}$ (eV)	$E_P^{(2)}$ (eV)	$E_O^{(3)}$ (meV)	SPIN$^{(4)}$ (10^{18}/cc)
CVD a-Si	550 C	1.44	3.396	11.69	70	2.4
	550 C in H_2	1.47	3.254	11.16	55	1.0
CVD a-Si	600 C	1.41	3.338	11.47	100	6.2
	600 C in H_2	1.43	3.303	11.45	100	4.6
CVD a-Si	650 C	1.40	3.262	10.99	90	10.5
	650 C in H_2	1.45	3.230	10.92	120	3.2
GD a-Si:H	250 C '81	1.63	3.393	11.40	50	-
	250 C '83	1.70	3.567	12.33	40	-
c-Si	-	-	3.924	12.79	10	-

(1) Tauc Gap from Eq.(9)
(2) Index dispersion parameters Eq. (10)
(3) Urbach Edge parameter at 300K
(4) Dangling bond spin density in film (g=2.00549-2.00587)

Table 2: Parameters for Eqs. (12) and (13) of text*

MATERIAL	θ_D (deg K)	θ_E (degK)	θ_E/θ_D	K (meV)	K_V (meV)	K_P (meV)
c-C	2200	1360	0.62	740	820	-80
c-Si	645	350	0.54	100	105	-5
a-Si:H	645	400	0.62	200	205	-5
c-GaP	446	277	0.62	115	119	-4
c-Ge	374	230	0.61	94	79	15
a-Ge:H	374	255	0.69	133	148	15
c-GaAs	344	213	0.62	92	58	34
c-InP	301	187	0.62	61	45	16
c-InAs	248	154	0.62	46	32	14

*Temperature dependence of optical gap taken from reference (7) except for: c-C ref.(28); c-Si ref.(22); a-Si:H ref(3); a-Ge:H ref(8).

CONTROL OF REACTION ON SUBSTRATE FOR PROPAGATION OF Si-NETWORK

Isamu SHIMIZU
The Graduate School at Nagatsuta, Tokyo Institute of Technology, Nagatsuta, Midori-ku, Yokohama, Japan 227

INTRODUCTION

Primary aim of this study is to understand the fundamental reactions on the substrate in making Si-network for the purpose of precluding the unfavourable natures in a-Si:H as follows:
1) instability under light illumination so called S-W effect,
2) poor hole-transport due to the deeply distributed tails at the valence band, and
3) low doping efficiency.

All these characteristics seems to be strongly related to structures of Si-network.

With regard to the deposition technology for the mass-production of the a-Si:H electronics devices, on the other hand, the control of the chemical reactions is essentially necessary to satify following requirements:
   a) higher growth rate,
   b) higher utility of the source gases, and
   c) higher quality.

RF glow discharge (RF-GD) of silane is considered to be one of the most promising preparation technique of a-Si:H and has extensively been applied to practical productions.[1] Relation between the Si-network and its defects have recently become of intense interest because all specific properties of a-Si:H are strongly related to the disorder of Si-network being far from the stable equilibrium states(non-equilibrium). Smith and Wagner[2] proposed an idear based on the concept of the thermal equilibrium that the defect density in a-Si:H is in an equilibrium with the strained energy storaged in film. Similar idea has been proposed by Street and his coworkers to elucidate the unexpected changes in electric properties for the doped a-Si:H at around 200 °C lower than the preparation temperature.[3] They gave a term of "the hydrogen glass" because the displacement of the local structure were mainly due to the diffusive nature of hydrogen through the Si-network. These concepts are very different from our feelings which we have had of the rigid tetrahedrally bonded Si-network.

These puzzles remind us of the topological arguments given by J.C.Phillips[4] in making Continuous Random Networks(CRN),in which the mean coordination number (C) is an important indicator. The glass-states are able to be formed only in the region of 2<C<3 since the optimum C value is 2.45. Accordingly, it is anticipated to be difficult to make the CRN with tetrahedrally bonded materials such as Si and Ge and some microstructures; voids or microvoids must be remained in the Si-network to release the strain. Large number of results done in early stage of a-Si and a-Ge gave a support to this idea until the first result illustrating the feasibility of valence control by doping small amount of impurities in hydrogenated amorphous silicon (a-Si:H) was reported by Spear and Lecomber[5]. Immediately after the paper was presented, efforts have been mainly devoted for the optimization of the preparation conditions under the hypothesis of that the RF-GD of silane was the best method for the preparation of a-Si:H.

Let us consider the fundamental processes in the RF-GD of silane in respect of the chemical reactions. Those are distinguished into two; i.e., the gas-phase reactions and the reactions in the visinity of the surface. In gas-phase, reactions for plasma-induced decomposition and the secondary reactions take place with the mother molecule(SiH_4), resulting in the precursors given in a form of $SiHn(n=0,1,2,$ and $3)$. Due to the strong chemical activity of these neutral radicals, polymeryzation and powder formation are often caused when the deposition velocity is forced to be increased by applying high RF power. Attempts have been made to make films in low pressure plasma, namely, ECR plasma[6] or simple microwave plasma so as to preclude the secondary reactions of the active species with the mother molecules. With regard to the qualities; electric and optical properties, further more efforts must be devoted to have a-Si:H with the quality being equivalent to that from RF-GD. In the preparation techniques under low pressure plasma, control of the reactions on the substrate is essential to make high quality films because all chemical species resulting from the plasma-induced decomposition may directly impinge on the substrate, which is the tradeoff to the enhancement in the utility of source gases and the preclusion of polymerization in the gas phase.

These facts have led us to investigate the chemical reactions on the substrate, which is anticipated to rule the structure of Si-network.

CHEMICAL REACTIONS FOR PROPAGATION OF Si-NETWORKS

We have brought our attentions into focus on the chemical reactions and the their control in the propagation of Si-network. Generally speaking, the reactions in the vicinity of the substrate consist of the two processes, i.e., sticking of the precursors and propagation of Si-network accompanying release of terminators, (H or F),. The sticking efficiency depends greatly on the precursors (SiH_n $0<n<3$)[7] and is almost indepent of the substrate temperature, implying that the neutral radicals stick on the substrate without any assists of kinetic energy. Similar behavior is observed in the precursors, SiH_nF_m ($n+m\leq3$), resulting from the plasma-induced decomposition of the gaseous mixture of SiF_4 and H_2.[8]

Measurements of the sticking efficiencies for various precursors have been made by different methods. Tanaka-Matsuda,[9] Perrin el.al.,[10] and Knights[11] obtained the values of 0.3, 0.1 and less than 10^{-3}, respectively, for the sticking coefficient of SiH_3. The other precursors such as SiH_2, SiH and Si have higher sticking coefficients, which is considered to be one of the origin in the deterioration of the quality. The average sticking coefficient of 0.63 was obtained by Tanaka and Matsuda[9] for these precursors under an optimum condition in making high quality films. Consequently, the qualities of a-Si:H , in the other words, the microstructures of Si-network are anticipated to be determined by the kinetics of the reactions in the vicinity of the growing surface.

The microscopic structures with respect to the chemical

bonds related to hydrogens are an another indicator exhibiting the quality of films. In the high quality a-Si:H prepared by RF-GD, SiH bond is predominant and appearance of (SiH$_2$)n means deterioration in the quality with an increase in dangling bonds. In addition, some morphological structures become apparent in the poor quality films. These evidences seems to give a support to the validity of the topological arguments, namely, the difficulty in making CRN of the rigid elements such as silicon. Although the content of hydrogen, (CH at.%) would be one of the most important parameters in determining the structure of Si-networks, there are no crucial relationship between the quality and the CH in the films prepared by RF-GD.[12] It is believed that the dangling bonds are capsuled with hydrogen but the number of the dangling bonds (10^{15} cm^{-3}) is far less than the CH of over 15 at%. The number of CH >15 at% is considered to be too much to release the strain storaged in films to form the CRN.

Under the assumption that hydrogen in films plays an important role either for reduction of dangling bonds or in propagation of Si-CRN, we tried to make films by introducing atomic hydrogen for the purpose of controlling the structure of Si-network.

In the conventional FR-GD of silane, CH was mainly determined by the substrate temperature as shown in Fig.1[13] plotted as a function of substrate temperature(Ts °C). No distinctive differences are brought into this relationship by changing the preparation conditions, i.e., deposition rate, flow rate of gases and dilution with Ar, He and H$_2$, despite marked changes are found in the qualities. With an increase in the Ts, the CH falls linearly and becomes almost zero at around Ts=500 °C. At the low Ts, polymer-like films containing large amount of hydrogen are obtained. Accordingly, the rigidity of Si-network increases with rising Ts up to the temperature of Ts=350 °C, accompanying obvious reduction in the content of hydrogen in films. When Ts is increased furthermore, the dangling bonds density rises rapidly as a consequence of evolution of hydrogen terminating the dangling bonds. Consequently, there are at least two processes of the reactions on the surface:

(1) propagation reaction of Si-network associating evolution of terminators, and
(2) evolution of hydrogen being independent of Si-network.

The optimum substrate temperature is, therefore, determined in the range of 200≤Ts (°C)≤350 for the preparation of high quality a-Si:H. These nature should be limited by the kinetics of the reactions in the vicinity of the surface and consequently, the optimum deposition rate is also determined by the substrate temperature. Several attemps have been made to accelerate the reaction kinetics for the propagation of Si-network by means of physical parameters; i.e., by illunination with ir light from the CO_2 laser on the surface[14], by impinging ions or excited molecules generated in the ECR plasma.[15] In those attempts, however, no marked progress have been observed yet in the quality of a-Si:H.

ROLES OF ATOMIC HYDROGEN AND FLUORINES

Selection of the chemical agents is another way to promote the surface chemical reactions. An important fact that microcrystalline silicon (μc-Si) can be formed by RF-GD from SiH_4 under the condition not far from the optimum for the preparation of a-Si:H reminds us of the feasibility in making more ordered Si-network. Dilution of SiH_4 with an excessive amount of hydogen plays an important role in making the μc-Si. Matsuda interpreted this in terms of "Hydrogen Coverage", viz., the surface covered with hydrogen may promote diffusion of silicon on the surface[13]

We believed another way of thinking about the role of hydrogen playing in the propagation of Si-network.[8] Atomic hydrogen is the key species to promote the reactions of the propagation of Si-network in the vicinity of the growing surface. The curves 2 in Fig.1 shows the CH values in films prepared at the same Ts but varying flow of atomic hydrogen introduced on the growing surface. Within the detective limit of IR absorption measurement, there is no detectable deutorium in a-Si:H prepared by RF-GD of SiH_4 with an assist of atomic deutorium, which implies that the reactions with the atomic deutorium take mainly place in the vicinity of the growing surface to release hydrogen in the film. The CH falls rapidly with an increase in the flow of atomic D and tends to saturate at about 8 (at %). Further reduction in CH is difficult so far as the deposition is done by rather low RF power condition.. Rigidity of Si-network is, therefore, considered to be increased with reduction in CH.

Remarkable influence of the atomic hydrogen have been observed on the propagation reactios for the films prepared from silicon-fluorides, i.e., SiF_4, SiF_2H_2. Curves 3 in Fig.1 show the behavior in the films prepared from SiF_4 under the flow of atomic hydrogen. SiF_nH_m (n+m=3) are likely candidates of the precursors.[8] The CH value decreases rapidly with increasing flow of atomic hydrogen up to 2 at% at the constant substrate temperature of 350 °C.

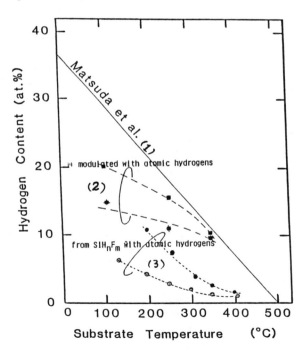

Fig. 1 The CH plotted as a function of Ts

Distinctive changes are found in the structures of Si-network associating with the reduction in CH. Namely, the amorphous Si-network is maintained up to the region of CH$\geq$ 5--6 at% and microcrystalline form (μc-Si) appears in CH of less than 5 at.%. Under the condision of introducing excessive amount of atomic hydrogen, μc-Si films are made at the deposition rate more than 20 A/s, which is quite different from the case from SiH$_4$. Accordingly, we can obtain both amorphous and crystalline films at the same deposition rate by controlling the flow of atomic hydrogen. This fact lends a support to our idea that atomic hydrogen offers a potential to promote the reactions for propagation of Si-network, and thus the term of (HR-CVD; Hydrogen Radical enhanced CVD) was given to the preparation techniques under control of atomic hydrogen.[16]

Webb and Veprek[17] elucidated the role of atomic hydrogen in making μc-Si in terms of chemical etching. However, preparation of μc-Si from SiF$_4$ is not the case because the etching velocity with atomic hydrogen is negligible compared with the deposition rate.

We verified the preferential evolution of fluorines from film despite F made a stronger bond with Si than hydrogen.

What is the role of fluorines playing in this preparation techniques? The precursors given in the form of SiF$_n$H$_m$ consisting of F and H and must carry F and H on the surface. Both the terminators, F and H, are evolved out from the film as a consequence of making the Si-network. The preferntial evolution of fluorines with the aid of hydrogen, therefore, may be interpreted in terms of the strong chemical interaction between F and H bound with silicon.

STRUCTURES OF Si-NETWORK

Measurements of Raman scattering spectra are one of the most powerful tool to investigate the structure of Si-network. According to the systematic studies made by Lannin[18], the Raman spectra are sensitive especially to the disorder of the bond angle. The width of the Raman peak observed at the wavenumber of about 480 cm^{-1} attributed to TO phonon is strongly related to the bond angle diorder. In Fig.2, Raman scattering spectra are shown for the films prepared from SiF4 with various CH by varying atomic hydrogen impinging on the surface. The width of the TO peaks decreases smoothly with reduction in CH, accompanying small shift of the peak position.

Similar behaviors are observed in the Raman bands(TO) of the films prepared from SiH4. The width becomes narrower as analogous manner with reduction in CH up to 5--6 at.%, and the sharp peaks appear abruptly at the wavenumbers of 510-- 520 cm^{-1} indicating appearance of crystalline phase as the CH is reduced less than 5 at%. The wavenumbers are continuously shifted toward higher energy with reduction in CH, implying that the volume fraction of μc-Si tends to increase with the reduction in the CH.

The appearance of the μc-Si was also verified by X-ray diffraction spectra. The introduction of atomic hydrogen is, therefore, considered to promote effectively the propagation reaction of Si-networks at low substrate temperature of around 200--350 °C.

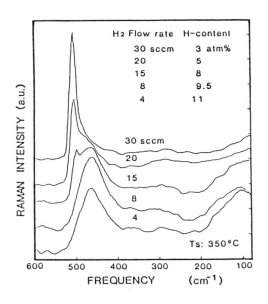

Fig. 2
Raman scatting spectra for Si thin films prepared under different flow rate of atomic hydrogens
The numbers indicate the flow and the content of hydrogen in the films

Under the condition being similar to make μc-Si on the glass, c-Si thin film is epitaxially grown on c-Si substrates, (110) and (100) at the temperature of 300 °C. It should be noteworthy that there is the optimum temperature to make the epitaxial growth at around 200--400 °C. The growth rate does scarecely affect to the structure of Si-network within the range of 20 A/s.

Different behaviors in the Raman scattering spectra were found of a-Si:H depending on the preparation techniques,[19] i.e., RF-sputtering and FR-GD. A remarkable differnce is also observed in the behavior of the films prepared with the aid of atomic hydrogen. Hishikawa[20] observed the behaviors in the films with the "device-quality" from silane that the wavenumber of the TO band shift to higher wavenumber with reduction in CH, implying an increase in the strain in the streching mode of bonds. In our films,the disorder in the bond angle is reduced with reducing CH with the aid of atomic hydrogen. Accordingly, rigidity of Si-network may increase with reducing CH up to 5--6 at.% by impinging atomic hydrogen without accompanying an increase in the internal stress because the films are prepared under a constant substrate temperature of 200--350 °C.

In early time, Knights[21] found some morphological structures in a-Si:H prepared by RF-GD of silane under certain conditions apart from the optimum one. The appearance of the structures was tightly related to the degradation of the quality. He and co-workers have successfully distinguished the conditions being optimum in making high quality a-Si:H from others by means of a sophisticated method termed "the trench coverage".[22] The walls in a narrow trench are smoothly covered with homogeneous a-Si:H film under the optimum condition (CVD-like), while inhomogeneity appears in the shape of the step-coverage with the morphology under the conditions aprt from the optimum (PVD-like). The large sticking coefficient of the precursors is anticipated to be responsible for these differences in the structures of the Si-network. However, μc-Si which has a more ordered structure is made under the condition of the PVD-like with the aid of the atomic hydrogen. This fact suports our idear that structural relaxation of the Si-network in the vicinity of surface is promoted greatly by the ability to move through the Si-network. Consequently, the atomic hydrogen is the effective tool to control the concentration of the hydrogen in the growing surface because of its strong chemical affinity in respect of Si-Si bond. In other words, the Si-Si bond is considered to be broken by making Si-H bond with atomic hydrogen, which is the reverse reaction with respect to the propagation reaction of Si-network. Accordingly, condition to form the ordered Si-network, crystallization, is made when the reactions in both directions will be equilibrated.

GROWTH KINETICS

The structure of the Si-network is primarily determined by kinetic factors, which is supported by the evidences that the chemical activities of the precursors and impinging of the atomic hydrogen give a great influence on the structures. The "hydrogen coverage" is another concept to illustrate controlling the kinetics related to the two dimentional surface diffusion of the precursors. Some evidences obtained in the fluorinated compounds are not the case at least because the reaction rate of the propagation is itself much faster than that from the hydrides. There must be some distinctive differences in the reaction kinetics between the film from the fluorinated precursors and from the hydrides.

Let us assume that the propagation reaction of Si-network takes place resulting from the release of terminators, H or F. The rate equation in respect of the CH is given by

$$dCH/dt = J - \{k + k'(F)\}CH \quad \text{---(1)}$$

Here, k, k' and (F) indicate the rate constants and the concentration of fluorines. J is proportional to the impinging rate of the precursors. Assuming the psedo-first order reaction under the hypothesis that the reaction is limited by the movable hydrogen being proportional to the CH, the eq.(1) is rewritten as follows;

$$dCH/dt = J - k''CH \quad \text{---------(2)}$$

Here k" indicates the rate constant. Under the steady-state

condition which is satisfied during deposition, the kate constant k" is giben by

$$k'' = J/CH \quad \text{---------------} \quad (3)$$
$$(k'' = k + k'(F))$$

In Fig.3, log (CH^{-1}) is plotted as a function of the inverse of the absolute temperature (Ts^{-1}) for films prepared from silane(a), SiF_4(b), and $SiCl_3H$(c), respectively. Marked differences are seen in the behaviors in these Arrhenius plots. Both the activation energies and the pre-factors obtained by the extraporation at the intercept with the ordinate in the (b) and (c) are much larger than (a). From these results, we can deduce the difference in the kinetics in making Si-network as follows:

(1) The small acivation energy and the small pre-factor in the (a) support that the propagation reaction of the Si-network is mainly promoted by the diffusion of the movable hydrogen through the Si-network.

(b) The strong chemical interaction between halogens, F and Cl, and hydrogen bound with silicon may be the origin of an increase in the pre-factor arisen from the change in the entropy during the transient states.

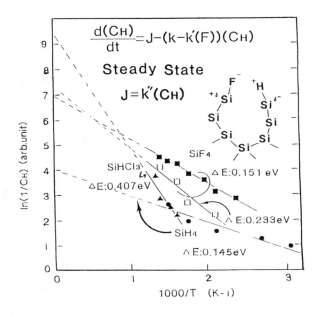

Fig.3 Arrhenius plots for the reactions in the vicinity of the growth surface
A schematic figure for the interaction (Coulombic) is illustrated in the inset.

The unexpected large activation energies attained in (b) and (c) is illustrated in terms of the three dimentional island-like growth because of the strong terminating behavior of the halogens bound with silicon. The chlorines are hard to be removed from films at low temperature because of its large radii comapring with that of F due to its steris hinderance.

Topological strain storaged at the growing surface is anticipated to be efficiently reduced by the excessive amount of monovalent elements, H, F, Cl. Accordingly, the progation reaction to make Si-CRN is effectively promoted by the freedom for the movements. Atomic hydrogen plays an important role in this reacting zone for supplying terminators made bonds with silicons, resulting in an increase in the freedom.

CRYSTAL GROWTH AT LOW TEMPERATURE

Under supplying excessive amount of atomic hydrogen, μc-Si was deposited as the same velocity to that for deposition of a-Si:H in a constant Ts of 200--350 °C. c-Si thin films were epitaxially grown on c-Si (110) and (100) under the similar condition to make μc-Si. No distinctive differences were found in the structures observed in Raman scattering spectra and RHEED when the deposition rate was changed from 5 A/s to 20 A/s. This fact implied that the kinetics is not limited the construction of Si-networks within this frameworks. By making condition to equlibrate the reactions for propagation and breaking the Si-Si bond with control of atomic hydrogen, the temperature for the two dimentional growth is effectivel reduced by 300 °C or lower.

CONCLUSIVE REMARK

Attempts have been made to review the preparation techniques of a-Si:H with regard to the chemical reactions on the substrate for the propagation of Si-network. The microstructures of the Si-network are considered to be determined mainly by the kinetic factors of the reactions on the substrate.

Impinging with atomic hydrogen on the substrate accelerates the propagation reactions and results in the rigid Si-network. The existence of fluorines enhances greatly the propagation reactions due to the long distance chemical interactions(Coulombic interaction) between H anf F bound with silicon. These and those lead us to an idea that reaction kinetics are enable to be controlled by atomic hydrogen and fluorine in the reacting zone. Sufficient condition is provided by this idea in making c-Si grown epitaxially at the substrate temperature of 200--350 °C. We conclude a novel preparation technique termed HR-CVD offers potential way to improve either quality or the production of a-Si:H and related materials.

REFERENCES
1. "Semiconductors and Semimetals" ed. by.J.I.Pankove, 21 A-D(1984)
2. Z E.Smith and S.Wagner, Phys.Rev.B32 (1985)5510
3. R.A.Street,J.Kakalios,C.C.Tsai and Hynes; Phys.Rev. Lett.,B35 (1987)1316
4. J.C.Phillips; J.Non-Cryst.Solids 35&36(1980)1157
5. W.E.Spear and P.G.LeComber;Solid State Commun.17(1975)1193
6. T.Watanabe, K.Azuma,M.Nakatani,K.Suzuki,T.Sonobe,and T.Shimada;Jpn.J.Appl.Phys.,25(1986)1805
7. J.Schmitt;J.Non-Cryst.Solids,59&60(1983)649
8. N.Shibata,K.Fukuda,H.Ohtoshi,J.Hanna,S.Oda,and I.Shimizu;Mat.Res.Soc.Symp Proc.95(1987)225
9. A.Matsuda and K.Tanaka;J.Appl.Phys.,60(1986)2351
10. J.Perrin,T.Broekhuizen and R.Benfehrat; European Mat.Res.Soc. Proc. Strassburg(1986)
11. J.C.Knights; Mat Res.Soc. Proc.,Boston(1986)
12. G.D.Cody;"Semiconductors and Semimetals",21B(1984)11
13. K.Tanaka and A.Matsuda; Mat.Res.Rep.,2(1987)139
14. N.Fukuda,K.Miyachi,H.Tanaka,T.Igarashi and S.Yamamoto ;Mat.Res.Soc.Symp.Proc.70(1986)25
15. M.Kitagawa,S.Ishihara,K.Setsune,Y.Yamabe, and T.Hirano; Jpn.J.Appl.Phys.,26(1987)L231
16 J.Hanna,N.Shibata,K.Fukuda,H.Ohtoshi,S.Oda, and I.Shimizu; "Disordered Materials", ed.by M.Kastner,G.A.Thomas, and S.R.Ovshinsky,(Plenum.1987)pp435
17. A.P.Webb and S.Veprek; Chem.Phys.Lett.,62(1979)173

18. J. S.Lannin; J.Non-Cryst.Solids,98&97(1987)39
19. R.Tsu; J.Non-Cryst.Solids,97&98(1987)163
20. Hishikawa; J.Appl.Phys,62 (1987)3150
21. J.C.Knights; J.Appl.Phys.,18Suppl.1(1979)101
22. C.C.Tsai,J.C.Knights,G.Chang and B.Becker;J.Appl. Phys.,59(1986)2998

CHANGES IN GAP-STATE PROFILES OF OF P-DOPED a-Si:H INDUCED BY LIGHT SOAKING AND THERMAL QUENCHING

H. Okushi, T. Furui,[a] R. Banerjee [b] and K.Tanaka

Electrotechnical Laboratory
1-1-4 Umezono Tsukuba-shi Ibaraki 305, Japan

ABSTRACT

Systematic studies on thermally and light-induced changes in the gap-state profiles of P-doped a-Si:H were performed using isothermal capacitance transient spectroscopy (ICTS). Two different features are found out in photo-induced changes; (1) gap states (N_{CT}) were created in the range of 0.25-0.35 eV below the conduction band edge E_C, which are attributed to ^{31}P-related-hyperfine-ESR centers, and (2) the conversion from deeper defect states located 1.0-1.2 eV below E_C (*D$^-$) to shallower defect located 0.4-0.6 eV below E_C (D$^-$). On the other hand, no suggestive changes in the density-of-states distribution at least in the energy range 0.25-1.5 eV below E_C were observed after thermal quenching, although shallow states above Fermi level E_F increased. It suggests that thermally and light-induced effects are independent of each other from the viewpoints of defect creation.

1. INTRODUCTION

Light-induced metastable changes[1-3] and more recently thermally induced metastable changes[4-6] in the electronic properties of hydrogenated amorphous silicon has evoked considerable interest because these metastable changes are directly associated with the stability of amorphous silicon based photosensitive devices. Although many works have been done for elucidating the mechanism of these changes, in particular of the light-induced metastable changes (Staebler-Wronski effects),[1] physics underlying those phenomena has not yet been well understood.[2,3]

The question of much interest is regarding origin of thermally and light induced changes - whether they can be ascribed to the same source or not? Reports by Smith et al.[4] for undoped a-Si:H and by Kakalios and Street [5] for doped a-Si:H suggest that the defect centers created by light and those formed during thermal equilibration process are indentical. Stutzmann, on the other hand, has shown that, for doped a-Si:H, effects of light soaking and thermal quenching are independent of each other.[6]

Several reasons can be raised why a real origin of those metastable changes has not yet been made clear up to now. However, one main reason should be a lack of exact information on the gap states profile of a-Si:H. Since we proposed an application of the isothermal capacitance transient spectroscopy (ICTS) method to study gap states in a-Si:H,[7] we have developed several variations of ICTS technique (dark and photo-ICTS)[8,9] and have done a systematic work on the gap states distribution in P-doped a-Si:H.[10,11] As a result, quite recently, a precise energy profile of localized states has been determined in the range of 0.25 eV < IE-Ecl < 1.6 eV. We applied this technique to the study of thermally and light-induced changes in the gap states of P-doped a-Si:H,[8,12-14] which will be described in this paper.

2. EXPERIMENTAL

Specimens used in the ICTS measurements were prepared in a 13.56 MHz inductively-coupled plasma chamber at low power density (r.f., voltage 250 V) from the PH_3-SiH_4 gas mixture. A flow rate of 5 SCCM, a gas pressure of 50 mTorr and a substrate temperature 300°C were maintained during the deposition of the film. The optical gap E_O determined by the *Tauc plot.* and the activation energy of dark conductivity of each specimen are around 1.7 eV, and 0.25-0.34 eV, respectively. Schottky barrier diodes of the above P-doped a-Si:H were fabricated in a sandwich configuration with n crystalline Si ($\rho \cong 0.01$ Ωcm) and thin semi-transparent Au metal contact with a diode area of 1 mm^2. Light soaking was done by exposure to AM1 light (100 mW/cm^2) for 4 hours. For thermal quenching, samples were annealed at 200 °C in a flow of hydrogen for 30 min and subsequently transferred to a liquid-nitrogen-cooled cryostat.

3. RESULTS AND DISCUSSION
3.1 Gap-State Profiles Before and After Light Soaking

Figure 1 shows the temperature dependence of dark ICTS spectra before ((a): annealed state) and after light soaking ((b): soaked state) in P-doped ($PH_3/SiH_4 = 1.5 \times 10^{-4}$) a-Si:H Schottky barrier.[12] The ICTS spectrum $S(t)$ is directly proportional to the gap-state density $g(E)$ as

$$S(t) = -(kTB) g(E(t)), \quad (1)$$

where $B = q\varepsilon A^2/2(V_D + V_R)$, q the electronic charge, ε the dielectric constant of the material, A the junction area, V_D the effective diffusion voltage at the junction, and V_R the reverse bias voltage. On the other hand, the time t along the abscissa corresponds to the inverse of the electron-emission rate (e_n). Then the time scale can be converted to the energy scale through the relation described as

$$E(t) - E_c = kT \ln\{v_n t\}, \quad (2)$$

where v_n is the attempt-to-escape frequency of thermal emission rate. A well-known expression derived from detailed balance arguments relates v_n to more fundamental parameters,

$$v_n = \sigma_n v_{th} N_c, \quad (3)$$

where σ_n is the electron-capture cross section of the defect, v_{th} the average thermal velocity of conduction electrons, and N_c the effective density of states in the conduction band,. In general speaking, σ_n is related to the nature of each trap level and both σ_n and v_n should depend on the energy

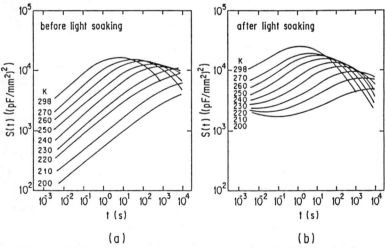

Figure 1. Temperature dependence of dark ICTS spectra before ((a): annealed state) and after light soaking ((b):soaked state) in P-doped ($PH_3/SiH_4=1.5\times10^{-4}$) a-Si:H Schottky barrier.

depth of trap level and temperature.

As reported earlier,[11] the maximum of $S(t)$ observed at T = 240-297 K originates from the doubly-occupied dangling bond (D^-) located at 0.5-0.6 eV below E_C. The energy location of the maximum has been determined from Eq.(2) with $v_n = 10^7 - 10^9$ s^{-1} which has also been derived from Eq.(3) using the data of the voltage-pulse-width dependence of $S(t)$.

As shown in Fig.1, the shapes of $S(t)$ before and after light soaking at each identical temperature are different from each other. At T = 298 K, a maximum is observed in the $S(t)$ spectrum. It increases and its peak position shifts towards shorter time constant by light soaking. At T= 200 K, though a maximum structure in $S(t)$ is not observed, $S(t)$ itself increases drastically by light soaking. The energy location of this region (denoted as N_{cT}) has been determined by using Eq.(2) as 0.25-0.4 eV below E_C (; near the Fermi level E_F).

Stutzmann have reported the observation of the ^{31}P-related-hyperfine-ESR centers and the increase of these centers by light soaking.[6] This result has been confirmed by Yamasaki et al. using ^{1}H-ENDOR-detected ESR.[15] In general, the ESR centers are located below E_F or near E_F since the centers arise from unpaired electrons. Therefore, it may be concluded that the photo-induced gap states N_{cT} observed by dark ICTS at 200K originate from the defects contributing to the hyperfine ESR signal.

The information of the gap-state profile from the midgap down to the valence-band edge (E_V) has been obtained by several variations of photo-ICTS.[8,9]

Figure 2 shows two different modes of photo-ICTS ($S(t)_{ON}$ and $S(t)_{OFF}$) using band-gap excitation (He-Ne laser hv =1.96 eV with 300 μw/cm2) as well as a dark ICTS spectrum ($S(t)_{dark}$) in P-doped a-Si:H. In the figure, $S(t)_{ON}$ was obtained from the increasing capacitance $C(t)$ with t, but $S(t)_{OFF}$ from the decreasing $C(t)$ with t.[14]

In the light-ON mode, the increase of $C(t)$ with t indicates that the positive space-charge in the depletion region increases by light exposure. It means that variation of the electron occupation in the depletion region is mainly due to the capture of excess holes generated by optical excitation. As the light-OFF mode corresponds to the reverse process of the light-ON mode,

$S(t)_{OFF}$ represents the gap-state profile of those hole trapping centers related with $S(t)_{ON}$.

Figure 3 shows the intensity dependence of $S(t)_{OFF}$ spectra before and after light soaking using the same sample in Fig.2.[14] As shown in the figure, the peak position of $S(t)_{OFF}$ shifts to the short time with increasing of the light intensity. This corresponds to the shift of the quasi-Fermi level of

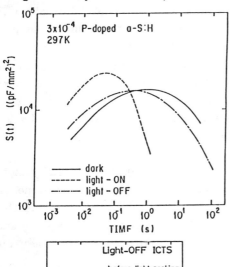

Figure 2. ICTS spectra using two different modes ($S(t)_{ON}$ and ($S(t)_{OFF}$) with band-gap excitation (Ne-Ne laser $h\nu$=1.96 eV with 300 μmW/cm^2) as well as a dark ICTS spectrum in P-doped (PH$_3$/SiH$_4$=3x10^{-4}) a-Si:H.

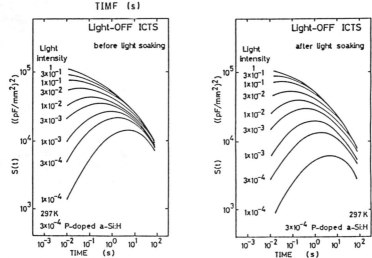

Figure 3. Relative light intensity dependence of light-OFF mode of the ICTS spectra of P-doped (PH$_3$/SiH$_4$=3x10^{-4}) a-Si:H. before and after light soaking. Note: relative intensity of 1x10^{-3} corresponds 300 μmW/cm^2).

excess holes. Therefore $S(t)_{OFF}$ in a time range longer than the time giving a $S(t)_{OFF}$ peak represents the actual density-of-state distribution of gap states, but not in a range shorter than the time. In the figure, the saturated $S(t)_{OFF}$ represents the gap-state profile at around 0.4-0.7 eV above E_V with the density of state of about 10^{18} cm^{-3}eV^{-1}, which was obtained using the hole capture cross section of $\sigma_p = 10^{-16}$ cm^2 determined from $S(t)_{ON}$.

Regarding the origin of the gap states located 0.4-0.7 eV above E_V (1.0-1.3 eV below E_C), we have suggested the doubly-occupied dangling bond coupled with P_4^+ (denoted as *D$^-$) on the basis of various experimental results such as ICTS,[10,16-18] PAS[19] and sub-bandgap PL.[20] Namely, in our model, at least two different sorts of D$^-$ states exist in the gap of P-doped a-Si:H. The concept of *D$^-$ is closely related with the doping scheme proposed by Street in terms of the pair creation of P_4^+ and D$^-$,[20] but in the present case, *D$^-$ requires a spatially intimate coupling of the pair, i.e., an intimate pair of P_4^+ and D$^-$ as shown in Fig.4.

This two defect model of D$^-$ and *D$^-$ was directly confirmed by photo-ICTS using below-gap excitations,[9] and supported by other experimental results. Several measurements such as ODMR,[21] CPM,[22] phase shift analysis of modulated photocurrent,[23,24] modulation spectroscopy,[25] and space charge limited current[26] and other carrier transport data[27] on

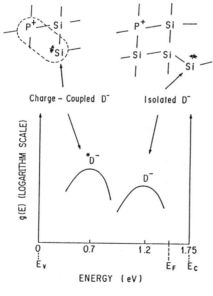

Figure 4. A schematic representation of the model of *D$^-$ and D$^-$.

undoped a-Si:H indicate that D⁻ is located at 0.5-0.6 eV below E_C. In doped a-Si:H many experimental results[22,25,28,29] indicate that D⁻ is located around 0.9-1.2 eV below E_C. Recently, Kocka[22] have explained the different energy location of D⁻ states by using the concept of dopant defect pairs which is essentially the same as that of our model of D⁻, *D⁻. Hirabayashi et al. observed the dopant-defect pairs in photo-induced-absorption detected ESR measurement in P-doped a-Si:H.[30]

As shown in Fig.3, the intensity $S(t)_{OFF}$ spectra decreases by light soaking, which means the density of gap states located at 1.0-1.2 eV below E_C (*D⁻) decreases by the light soaking. The behavior of *D⁻ is strongly correlated with that of D⁻ states located at 0.5-0.6 eV below E_C, namely, an increase of one is always accompanied by a decrease of the other. These complementary change between *D⁻ and D⁻ suggests that the conversion from *D⁻ and D⁻ occurs by light soaking and the reverse conversion is induced on subsequent room-temperature annealing.

On the other hand, the photo-ICTS spectra at low temperature (200 K < T < 250 K) indicate that the gap states located at 0.2-0.5 eV above E_V,

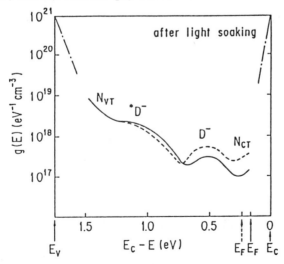

Figure 5. The gap-state profiles of P-doped a-Si:H before (solid curve) and after (dotted curve) light soaking, which were collected by measurements of dark and photo-ICTS in the temperature range 200-298 K.

denoted as N_{VT}, does not change significantly by light soaking. It is to be noted however that the results of the photo-ICTS are ambiguous due to an known factor of the electron occupation function of gap states under illumination, which causes an estimation error of 20 % in the density of gap states.

Figure 5 summarizes the present results on the gap-state profile of P-doped a-Si:H before and after light soaking. As shown in the figure, the change of the gap states by light soaking can be characterized by two main features; one is the $^*D^-$-D^- conversion and the other is the creation of N_{CT}. From these results, we conclude that the total density of dangling bonds, namely, the summation of the density of D^- and $^*D^-$, does not change essentially but the density of N_{CT} increases by light soaking.[12,14]

3.2 Thermally Induced Changes

Figure 6 shows the density-of-state distribution $g(E)$ of P-doped a-Si:H before and after thermal quenching where specimens were annealed at 200 °C in a flow of hydrogen for 30 min, and subsequently transferred to a

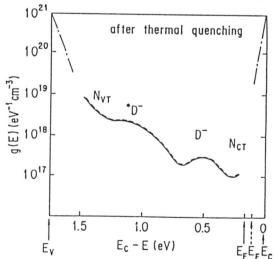

Figure 6. The gap-state profiles of P-doped a-Si:H before (solid) and after (dotted) thermal quenching, which were collected by measurements of dark and photo-ICTS in the temperature range 200-298 K.

special liquid-nitrogen-cooled cryostat.[13,14] As shown in the figure, no suggestive changes are observed either way, in the density of state near the conduction band tail (N_{CT}), D^- centers, $^*D^-$ centers, or near the valence band tail (N_{VT}), at least in the energy range from 0.25 eV to 1.50 eV below E_C. Although $g(E)$ determined by the photo-ICTS measurement has an ambiguity mentioned above, the difference between Fig.5 and Fig.6 is clear because the same experimental procedures were applied for both.

Thus, there is a clear difference in the effects between thermal quenching and light soaking as far as gap states profile is concerned. It should be noted that no significant change in $g(E)$ in the range 0.25-1.50 eV below E_C means no creation of dangling bonds by thermal quenching in the present P-doped a-Si:H specimens. This is consistent with the results of Street et al., where studies by PDS and C-V, have yielded for them no change in the dangling bond density.[28]

On the other hand, thermal quenching effects were clearly detected as changes in the junction capacitance C_0 or in the C-V characteristics of the Schottky barrier diodes used for the ICTS measurements as well as in the dark conductivity of the a-Si:H film; specifically, the C-V characteristics were

	Thermal Quenching	**Light Soaking**
C_0	Increase Then gradual decrease with room temperature annealing with bias	Increase Then gradual decrease with room temperature annealing with bias
N_{cT}	No change	No change
D^-	No change	Increase Then gradual decrease with room temperature annealing with bias
$^*D^-$	No change	Decrease Then gradual increase with room temperature annealing with bias
N_{vT}	Not discernable	Not discernable

Table 1. The changes in the gap-state profile and the junction capacitance after thermal quenching and light soaking.

changed and the junction capacitance C_0 increased drastically after the thermal quenching. For example, C_0 under a reverse voltage of 2 V became about 900 pF/mm^2 after thermal quenching, while C_0 under the same bias at rested states was about 500 pF/mm^2. The dark conductivity increased after thermal quenching as reported by Street et al.[31,32] These results mean that shallow donors and therefore free electrons increase by thermal quenching, indicating a creation of gap states located outside the present energy range (0.25-1.50 eV below E_c) covered by the ICTS measurements.

Thus, the D$^-$ center changes by light soaking but not by thermal quenching. Similar experiments concerning other gap states such as *D$^-$ and N$_{CT}$ also indicate that these states change by light soaking but not by thermal quenching. The changes in the gap-state profile and the junction capacitance both thermal quenching and light soaking are summarized in Table 1.

3.3 Sources of Thermally and Light-Induced Changes

Regarding problem whether thermally and light induced changes can be ascribed to the same source or not, the present results agree with Stutzmann, that the effects of light soaking and thermal quenching are independent of each other.[6] However, in some finer details, the conclusions are different. Stutzmann's model involves the increase in dangling bonds on light soaking, with the variety confined to D$^-$ centers. *D$^-$ centers have not been spoken of. ICTS experiments points to no change in the total dangling bond density, but conversion from *D$^-$ to D$^-$.

Secondly, Stutzmann[6] and also Yamasaki et al.[15] have found that light soaking in P-doped a-Si:H causes an increase in the ^{31}P-related-hyperfine-ESR hyperfine center and this has been attributed in the former case to normal neutral donors, P$_4^0$,[33] but in the latter to anti-bonding states of the weak Si-P bond occupied by electrons, i.e., P$_4^0$ with a weak Si-P bond.[17] As described above, the dark ICTS measurement at the low temperature T < 240 K has suggested that the ^{31}P-related-hyperfine-ESR centers correspond to the gap states N$_{CT}$ which increase by light soaking as shown in Fig.5.

According to Stutzmann, shallow donors of normal four-fold coordinated P atom (P$_4^+$) are created by light soaking since an increase in the density of ^{31}P-related-hyperfine-ESR centers would mean an increase in the density of normal neutral donors (P_4^0), and P_4^+ as well, in his model.[6] If the density of donors happens to increase, it would mean an upward shift in the Fermi level but in actuality the movement is in the opposite direction. Stutzmann has explained this by the creation of dangling bonds on light soaking.[6] However, the present results in Fig.5 indicate that there is no creation but only conversion of dangling bonds. Moreover this conversion is in a direction that would be compatible with an upward shift of the Fermi level. Thus, the assumption that the ^{31}P-related-hyperfine-ESR centers originate from the normal neutral donors is not compatible with the present experimental results.

This conclusion can also be confirmed from the experimental results on thermal quenching. As mentioned above, the results in Fig.6 indicate that the states (N_{CT}) located near E_F as well as deeper states (D^-, $^*D^-$ and N_{VT}) do not change at all by thermal quenching. However, the junction capacitance C_0 or the effective density of the space charge in the depletion region increases after thermal quenching, suggesting that an actual change in shallow states occurs by thermal quenching. Therefore, it is clear that origin of shallow gap states contributing to the change in C_0 is essentially different from that of the N_{cT} states located near E_F. Street et al, suggested that those shallow states created by thermal quenching originate from normal four-fold coordinated P atoms on the basis of sweep-out experiments.[28,29] Thus it seems that the N_{CT} states (i.e., the ^{31}P-related-hyperfine-ESR centers in our model) should not be attributed to the normal neutral donors (P_4^0),[33] but the ^{31}P-related-hyperfine ESR originates from the P-Si weak bond occupied by an electron (P_4^0 with a P-Si weak bond), which has been proposed by Tanaka et al.[17] and Ishii et al.[34]

3.4 Mechanism of Light Induced Change

Concerning the mechanism of the $^*D^-$-D^- conversion and the creation of the N_{CT} states which may come from ^{31}P-related-hyperfine ESR, our previous model becomes more realistic.[17] In this model, the transformation from $^*D^-$ and D^- and creation of Si-P weak bonds (N_{CT}) occur simultaneously through a strong electron-lattice coupling when the excess holes are captured by $^*D^-$. An excess energy from photo-excited carriers is released to the lattice system surrounding *D⁻ through a strong electron-lattice coupling. This might give a strong perturbation to the P_4^+-*D⁻ charge couple and make a conversion from *D⁻ to D⁻, namely, the charge coupled dangling bond states become to isolated states.

One possible interpretation for the conversion from *D⁻ to D⁻, for example, is that the H atom of H-Si bond located near *D⁻ kicked out by a strong perturbation of carrier capture process and moves toward the *D-atom and forms H-Si bond with the *D⁻ atom. This corresponds to annihilation of *D⁻ and creation of D⁻ as is illustrated in Fig.7 This conversion, with some probabilities, is accompanied by the creation of Si-P weak bonds depending on local toporogical constraint produced by a variety of structural randomness. An electron, once trapped at P_4^0 having a Si-P weak bond, is energetically stabilized at the anti-bonding state of the weak Si-P bond. Therefore, the P_4 states remain neutral (P_4^0) even at steady state of room temperature, causing a downward shift of E_F. However, if the conversion is not accompanied by the creation of Si-P weak bonds, the P_4 states are positive (P_4^+) at steady state and D⁻ restores original *D⁻ by thermal annealing at room temperature because of existence of Coulomb interaction between P_4^+ and D⁻ and the relaxation of local topological constraint.

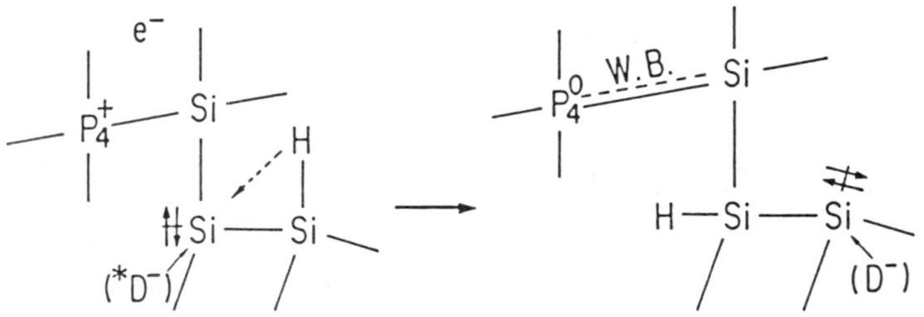

$$P_4^+ + {}^*D^- + e^- \longrightarrow P_4^0 \text{ with a weak bond} + D^-$$

Figure 7. A schematic diagram of *D⁻ - D⁻ conversion accompanying the formation of P with a Si-P weak bond.

3.5 Mechanism of Thermally Induced Change

Concerning the thermally induced effect, some experimental results in the present study need further clarification. However, the present results strongly suggest that the thermal induced effect is not associated with deeper gap states such as D^- and $^*D^-$ but with shallower gap states of P_4^+. Therefore, we emphasize that, if the thermally induced effect is due to the creation and annihilation of the donor states ($P_3^0 \leftrightarrow P_4^+ + e^-$) as pointed out by many authors,[5,6,31,32] one must consider the creation of donor states without the creation of dangling bonds that seems to be unlikely.

4. SUMMARY

Systematic studies on thermally and light-induced changes of the gap-state profiles in P-doped a-Si:H were performed using several variations of the ICTS technique. Two different features were found out in photo-induced changes; (1) gap states (N_{CT}) are created in the energy range of 0.25-0.35 eV below E_C and have been attributed to ^{31}P-related-hyperfine – ESR centers, (2) defect conversion takes place, i.e., $^*D^-$ coupled with P_4^+ in the range of 1.0-1.1 eV below E_C is transformed to normal D^- located at around 0.4-0.6 eV below E_C.

In case of thermal quenching, on the other hand, no suggestive change was observed either way, in the density of states near the conduction band tail (N_{CT}), the deep gap states (D^-, $^*D^-$), or near the valence band tail (N_{VT}), at least in the energy range from 0.25 eV to 1.50 eV below E_C.

From these result, changes in the gap-state profile of P-doped a-Si:H by light soaking and thermal quenching are different from each other, probably indicating that both effects are independent from the viewpoint of defect creation. However, no new dangling bonds seem to be created in both cases.

---

a) Permanent address: Tokai Univ., 117 Kitakaname, Hiratsuka-shi, Kanagawa 259-12, Japan.
b) Permanent address: Indian Association for the Cultivation of Science, Jadvur, Calcutta 700-032, India.

REFERENCES

1) Staebler,P.L. and Wronski,C.R., Appl. Phys. Lett. <u>31</u>, 292 (1977).
2) Wronski,C.R., in"Semiconductors and Semimetals" ed. by J.I.Pankove, <u>21C</u> (Academic Press, New York, 1984) p.347.
3) Fritzsche, in "Optical Effects in Amorphous Semiconductors " ed. by P.C.Taylor and S.G.Bishop, AIP Conf. Proc. <u>120</u> (New York, 1984) p.478; in "Stability of Amorphous Silicon Alloy Materials and Devices" ed. by B.L.Stafford and E.Sabisky, AIP Conf. Proc. <u>157</u> (New York, 1987) p.366.
4) Smith, Z E., Aljishi,S., Slobodin,D., Chu,V. and Wagner,S., Phys. Rev. Lett. <u>35</u>, 1316 (1987).
5) Kakolios,J. and Street,R.A., in "Stability of Amorphous Silicon Materiels and Devices ed. by B.L.Stafford and E.Sabisky, AIP Conf. Proc. <u>157</u> (New York, 1987) p.179.
6) Stutzmann,M., Phys Rev. <u>B35</u>, 9735 (1975).
7) Okushi,H., Tokumaru,Y., Yamasaki,S., Oheda,H. and Tanaka,K., Jpn. J. Appl. Phys. <u>20</u>, L549 (1981).
8) Okushi,H., Itoh,M., Okuno,T., Hosokawa.Y., Yamasaki.S. and Tanaka,K., in "Optical Effects in Amorphous Semiconductors" ed. P.C.Taylor and S.G.Bishop , AIP Conf. Proc. <u>120</u> (New York, 1984) p.250.
9) Okushi,H. and Tanaka,K., Phil. Mag. Lett. <u>B55</u>, 135 (1987).
10) Tanaka,K. and Okushi,H., J. Non-Cryst. Solids <u>66</u>, 205 (1984).
11) Okushi,H., Phil. Mag. <u>B52</u>, 33 (1985).
12) Okushi,H., Furui,T., Banerjee,R. and Tanaka,K., submitted to Appl. Phys. Lett.
13) Banerjee,R., Furui,T., Okushi,H. and Tanaka,K., Appl. Phys. Lett. <u>53</u>,1829 (1988)
14) Okushi,H., Furui,T., Banerjee,R. and Tanaka,K., to be published in Advances in Amorphous Semiconductors, H.Fritzsche, ed. (World Scientific, 1988).
15) Yamasaki,S., Kuroda,S. and Tanaka,K., in "Stability of Amorphous Silicon Alloy Materials and Devices" ed. B.L.Stafford and E.Sabisky, AIP Conf. Proc. <u>157</u> (New York,1987) p.197.

16) Okushi,H., Asano,A., Miyakawa,M., Yamasaki,S., Oheda,H and Tanaka,K., J. Non-Cryst. Solids 59/60, 393 (1983).
17) Tanaka,K., Okushi,H. and Yamasaki,S., in "Tetrahederally-bonded Amorphous Semiconductors" ed. D.Adler and H.Fritzsche (Plenum Publishing Corporation, 1985) p.239.
18) Yamasaki,S., Phil. Mag. B56, 79 (1987).
19) Tajima,M., Okushi,H., Yamasaki,S. and Tanaka,K., Phys. Rev. B33 8522 (1986).
20) Street,R.A., Phys. Rev. Lett. 49, 1187 (1982).
21) Morigaki,K., Sano,Y. and Hirabayashi,I., Solid States Commun. 39, 969 (1981).
22) Kocka,J., J. Non-Cryst. Solids, 90, 91 (1986).
23) Oheda,H., Yamasaki,Y., Yoshida,T., Matsuda,A., Okushi,H. and Tanaka,K., Jpn. J. Appl. Phys. 21, L440 (1982).
24) Nitta,Y., Abe,K., Hattori,K., Okamoto,H. and Hamakawa,Y., J. Non-Cryst. Solids 97/98, 695 (1987).
25) Vardeny,Z., Zhou,T.X., Stoddart,H. and Tauc,J., Solid State Commun. 65, 1049 (1987).
26) Schauer,F. and Kocka,J., Phil. Mag. B52, L52 (1985).
27) Spear,W.E., Steermer,H.L., LeComber,P.G. and Gibson,R.A., Phil. Mag, B50, L33 (1984).
28) Lang, D.V., Cohen,J.D. and Harbison,J.P., Phys. Rev. B25, 5285 (1982).
29) Street,R.A., Biegelsen,D.K., Jackson,W.B., Johnson,N.M. and Stutzmann,M., Phil. Mag. B52, 235 (1985).
30) Hirabayashi,I., Morigaki,K., Yamasaki,S. and Tanaka,K., in "Optical Effects in Amorphous Semiconductors" ed. P.C.Taylor, and S.G.Bishop, AIP Proc. 120 (New York, 1984).
31) Street,R.A., Hack,M. and Jackson,W.B., Phys. Rev. B37, 4209 (1988).
32) Street,R.A., Kakalios,J., Tsai,C.C. and Hayes,T.M., Phys. Rev. B35, 1316 (1987).
33) Stutzmann,M. and Street,R.A., Phys. Rev. Lett. 54, 1836 (1985).
34) Ishii,N., Kumeda,M. and Shimizu,T., Solid State Commun. 53 543 (1985)

SUBBAND OPTICAL TRANSITION IN AMORPHOUS SILICON QUANTUM WELL SYSTEMS

Kiminori HATTORI, Hiroaki OKAMOTO, and Yoshihiro HAMAKAWA

Faculty of Engineering Science, Osaka University
Toyonaka, Osaka, 560 Japan

ABSTRACT

Optical transitions between two-dimensional (2D) subband states in hydrogenated amorphous silicon (a-Si:H) quantum well (QW) structures have been extensively investigated by using differential absorption technique. The temperature-derivatives of absorption spectrum clearly show a staircase form for a-Si:H QW structures of the well width below 50A. These results suggest that 3D-parabolic band splits into a series of subbands. Analysis of the observed energies of subband transitions in terms of a square QW model leads to a conclusion that the electron and hole effective masses are $0.3\ m_0$ and m_0, where m_0 denotes free electron mass.

1. INTRODUCTION

Quantum well (QW) effects are well known in ultrathin layered structures of crystalline semiconductors, such as $Al_xGa_{1-x}As$ family. The most profound consequence of quantum confinement is the splitting of band structure into a series of two-dimensional (2D) subbands. The similar phenomena could be expected to be observable even for amorphous semiconductors so far as the disorder-induced broadening is less extensive as compared with the energy separation between the quantum levels. To study the band structure of amorphous semiconductor QWs is, therefore, of fundamental interest for understandings of the quantization effects as well as disordered system itself.

The most direct way for this purpose is to measure the optical absorption spectrum. The observed absorption spectrum of the QW structures,[1] however, does not exhibit any features distinguishable from that of the bulk materials, and hence provides us no definitive information about subband structures. The observation is likely inconsistent with the experimental result on resonant tunneling in amorphous semiconductor QW structure [2], which indicates the formation of subband states. The inconsistency may be removed if the structureless spectrum is ascribed to the non-direct optical transition process being operative in amorphous materials, which tends to reduce structures in the spectrum reflecting the quantized density of states. It is then required to introduce experimental approaches with high detection sensitivity for extracting these structures from the unresolved absorption spectrum. Such extreme sensitivity is considered to be obtained in differential (modulation) techniques [3].

In this paper, we present a simple model of optical absorption in amorphous semiconductor QW with assuming a completely broken momentum selection rule (Section 2), and experimental results of differential absorption spectroscopy on hydrogenated amorphous silicon (a-Si:H) QW structures (Section 3). The differential spectra measured for the QW layer of width less than around 50A exhibits a staircase behavior, which is interpreted as due to the subband transitions in accordance with the model. The observed behaviors are also reexamined by more realistic calculation, leading us to further understandings of the amorphous semiconductor QW (Section 4), and the plausibility of the interpretation is discussed.

2. A MODEL FOR OPTICAL ABSORPTION IN QW

Before going into the description of the experimental result on differential spectroscopy, we will briefly discuss what kind of absorption spectrum is expected for the interband transition in both bulk and quantum wells. It would be reasonable to start with Tauc's model for optical absorption [4], because it leads to quantitative predictions which are proved to be in good agreement with experiment at least for bulk material. The absorption coefficient α associated with the transition between 3D-parabolic bands in amorphous semiconductors is expressed by

$$\hbar\omega\alpha \propto (\hbar\omega-E_0)^2 U(\hbar\omega-E_0), \qquad (1)$$

within the context of the model assuming a completely relaxed selection rule for the wave vector k and a constant momentum matrix element. [4] Here E_0 denotes the Tauc optical gap, $\hbar\omega$ photon energy and U(E) a step function. If the dipole matrix element is assumed to be constant, the left-hand side of the relation (1) is replaced by $\alpha/\hbar\omega$. [4] However, this does not essentially alter the conclusion reached in the following discussion.

In an amorphous semiconductor QW structure, suppose that the well layer thickness is reduced to less than the carrier mean free path so that the carrier system in the well layer enters into the quantum regime. Quantization of the wave function in the direction perpendicular to the well layer plane brings about the subband states of constant density $m^*/(\pi\hbar^2 L_W)$, as schematically shown in Figure 1. The quantity m^* denotes the effective mass, and L_W the thickness of well-layer. Each subband is labeled by the confined state quantum number n (=1,2,3,...). Transitions between subband states should obey an approximate selection rule for confined states (Δn=0), whereas the k-selection rule in the direction parallel to the well plane remains relaxed. [5] Therefore, the absorption coefficient for transitions between subband states will follow

$$\hbar\omega\alpha \propto \Sigma_n (\hbar\omega-E_n) U(\hbar\omega-E_n), \quad n=1,2,3,\ldots, \qquad (2)$$

where E_n denotes energy separation between the n-th subband edges in the conduction and valence bands. Equation (2) indicates that the quantization effect does not produce a staircase structure as does in quantum wells based upon crystalline direct-gap semiconductors, instead only changes in the slope are induced at each transition energy E_n.

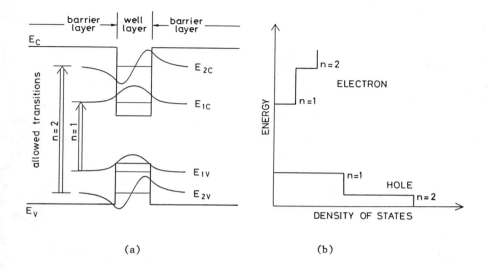

FIG. 1. QW structure. (a) Energy band diagram, quantum energy levels, and wave functions for the confined states. E_{1C} and E_{2C} are n=1 and n=2 electron levels, and E_{1V} and E_{2V} are n=1 and n=2 hole levels, respectively. (b) Density of states as a function of energy.

The spectral shapes of $\hbar\omega\alpha$ for both 3D-parabolic band transition in bulk material and 2D-subband transitions in QW structure are illustrated in Figure 2. As suggested in this figure, it is not an easy task to directly distinguish the absorption spectrum in the quantum regime from that in the unquantization regime, since the optical structures associated with the quantization effect may be subject to disorder-induced broadening.

The difference between the spectral shapes for the bulk and QWs appears to be more distinguishable in their derivatives with respect to either temperature or wavelength. The temperature(Θ)-derivative of absorption coefficient is given by

$$d\alpha/d\Theta = \Sigma_m (\partial\alpha/\partial E_m)(dE_m/d\Theta), \quad m=0,1,2,...,$$

yielding

$$\hbar\omega d\alpha/d\Theta \propto (\hbar\omega-E_0)U(\hbar\omega-E_0), \quad \text{for bulk,} \quad (3)$$

$$\propto \sum_n U(\hbar\omega-E_n), \quad n=1,2,3,\ldots, \quad \text{for QW.} \quad (4)$$

Equation (4) will be valid so far as the thermally induced changes in width and potential depth of the well are practically negligible so that $dE_n/d\Theta$ is identical for every subband index n. Important implications of equations (3) and (4) are that the derivative spectra for the QW structure exhibits a staircase form with steps at photon energies $\hbar\omega=E_n$, while those for unquantization regime have a linear dependence on photon energy.

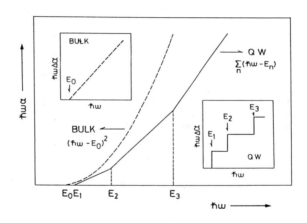

FIG. 2. Schematic model of $\hbar\omega\alpha$ spectra for 3D-parabolic band transition in bulk and 2D-subband transitions in QW. Insets show their derivatives $\hbar\omega\Delta\alpha$ with respect to temperature.

3. DIFFERENTIAL ABSORPTION SPECTROSCOPY ON a-Si QW

In this section, we will focus on photothermal modulation (PTM) spectroscopy to study interband optical transitions in a-Si:H/a-SiC:H multilayered structures.[6] The multilayers were prepared on SiO_2 glass substrate by rf plasma chemical vapor deposition using a 1:9 SiH_4/H_2 gas mixture for a-Si:H layers and 1:13:126 $SiH_4/CH_4/H_2$ gas mixture for a-SiC:H layers. Each layer was formed in separate chambers with interruption on the plasma to minimize interlayer cross-contamination. Substrate temperature was kept at 300°C during the growth. The a-Si:H well layer thickness was varied from 20A to 1000A while keeping the total well layer thickness at 3000A, and the a-SiC:H barrier layer thickness was fixed at 100A to avoid interlayer carrier tunneling. The optical band gaps were 1.75eV and 2.80eV for a-Si:H and a-SiC:H, respectively, which were determined from the optical absorption spectra of identical thick films following Tauc plot. The band discontinuities were estimated to be 0.80eV for the conduction band and 0.25eV for the valence band from the results of ultraviolet photoelectron emission measurement by using a Fowler plot. The multilayers are confirmed to be formed with abrupt heterointerfaces within an atomic scale from the x-ray diffraction measurement [7], of which the expected energy band diagram is illustrated in Figure 3.

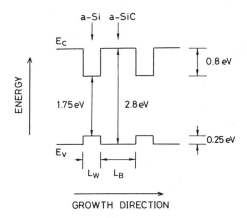

FIG. 3. Energy band diagram for a-Si:H/a-SiC:H multilayer.

The PTM spectroscopy is a conventional thermal modulation spectroscopy in which the temperature of the sample is modulated by heat produced by light absorption. The pump light from an Ar^+ ion laser (488nm) with an intensity of $1W/cm^2$ was mechanically chopped at 5Hz. White light with an intensity of $5mW/cm^2$ was used as a probe light for measuring the induced change in absorption. The transmitted light was dispersed by a monochromator, and the transmission T and its modulated component ΔT were detected by using a photomultiplier and a lock-in amplifier. The temperature-derivative of absorption coefficient is proportional to $S \equiv -\Delta T/T$ over the spectral region of photon energy above the absorption edge of the a-Si:H sample, as indicated by Pfost et al.[8] The samples were held in a cryostat, and measurements were performed at temperature of 100K to minimize thermal broadening of the PTM spectrum.

Figure 4 presents the PTM spectra $\hbar\omega S$ measured on the multilayered structures with the a-Si:H well layer thickness L_W=20,30,50 and 500A, as a function of photon energy $\hbar\omega$. Interference fringes which usually perturb the PTM spectrum in the lower photon energy side have been averaged out. For L_W=500A, it is clearly found that the spectrum has a linear dependence on the photon energy. Similar spectra have been obtained for multilayered structures of $L_W \geq 100A$, as well as in unlayered (bulk) a-Si:H films. This behavior is in good agreement with that predicted for the unquantization regime. We have confirmed that the threshold energy E_0 determined with eqn. (3) is completely equivalent to that determined by the Tauc plot, according to equation (1), over a temperature range from 100K to 300K.

The spectral shapes of the PTM signal for $L_W \leq 50A$ are entirely different from that for L_W=500A. These spectra clearly exhibit the staircase behavior expected from eqn. (4). The energy positions of step edges are indicated by arrows in the figure. As found for the case of L_W=50A, the energy interval between the step edges becomes longer at higher photon energy, indicating that the optical interference effect is excluded for the explanation. It may be, therefore, plausible to identify the staircase behavior as originating from subband transitions, so that the energy position of each step edge is related to the threshold energy E_n for the transition. If these assignments are correct, then the relative magnitude of the signal at each suggested threshold energy E_n should follow $E_n S(E_n)/(E_1 S(E_1))=n$, as predicted from eqn. (4). This relation can be readily confirmed in the experimental data in Fig. 4, suggesting the plausibility of the present assignments for threshold energies. On the other hand, the deviation from an ideal staircase form, that is, the spectrum broadening, may be associated with spatial fluctuations of the a-Si:H well layer thickness on an atomic scale, as well as the contribution from transitions involving diffused tail states which may exist below each subband edge. In addition, the effect of collisions may be included as a possible mechanism for the spectrum broadening.

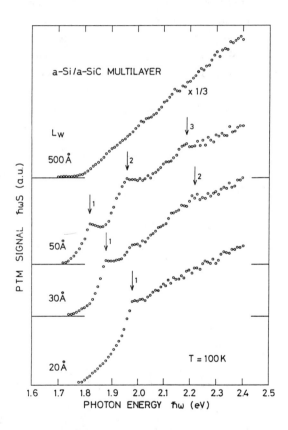

FIG. 4. PTM spectra $\hbar\omega S$ measured on the a-Si:H/a-SiC:H multilayers with the well-layer thicknesses L_W=20,30,50, and 500Å at temperature 100K. The arrows indicate the energy positions of step edges.

In our recent work on in-plane carrier transport in the a-Si:H/a-SiC:H multilayers by using transient grating method,[9] an anomalous increase in the carrier diffusion length was observed when the a-Si:H well layer thickness L_W was decreased below 50A. The result led us to conclude that the carrier transport takes place at the subband states in the a-Si:H QW for $L_W \leq 50A$. The present observation of subband transitions for $L_W \leq 50A$ is consistent with the conclusion reached by the transport experiment.

The threshold energies E_n for the transition between the n-th subband states are summarized in Figure 5 as a function of the well layer thickness L_W. The threshold energies are easily calculated on the basis of the square QW model. The result of calculation fitted to the experimental data are plotted by solid lines. Here the electron (m_e^*) and hole (m_h^*) effective masses are chosen 0.3 m_0 and m_0, respectively, in both sublayer regions, where m_0 denotes the free

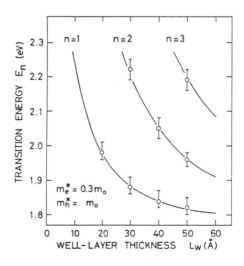

FIG. 5. Subband transition energies E_n (n=1,2,3) of the a-Si:H/a-SiC:H multilayers as functions of the well-layer thickness L_W. Circles are experimental data, and solid lines the theoretical plots calculated from the square QW model.

electron mass. As found in the figure, the theoretical plots yield good fits to the experimental data. Identical values of effective masses are deduced from wavelength-differential absorption measurement on a-Si:H single QW layers [10], whereas a larger electron mass ($m_e^* = 0.6 m_0$) is suggested by the analysis of the resonant tunneling phenomenon in a-Si:H/a-Si$_3$N$_4$:H double barrier structures [2]. This disagreement may imply that the electron mass in the a-SiN:H barrier layer takes a larger value than that in the a-Si:H layer.

4. SELF-CONSISTENT CALCULATION OF ABSORPTION SPECTRUM

The agreement between the spectral dependence expected from eqn. (4) and experimental data on the differential absorption of a-Si:H based QW structures confirms the physical insight contained in the simple model described in section 2. The model assumes a complete removal of the selection rule for the parallel k vector due to disorder, whereas the same disorder is assumed to have no profound effects on the quantized density of states. In this sense, the model is likely to involve essential inconsistency. To test the plausibility of the assumptions made in the model as well as the consequences derived from it, we have carried out a self-consistent calculation of the absorption spectrum in amorphous semiconductor QW structures on the basis of Abe-Toyozawa site-disordered model [11].

The theoretical calculation is made on a simple two-band system described by the tight-binding Hamiltonian of the form [11]

$$H^\mu = \Sigma_i |i\mu> E_i^\mu <i\mu| + \Sigma_{i \neq j} |i\mu> V_{ij}^\mu <j\mu|, \quad \mu = c \text{ or } v. \quad (5)$$

Here, $|i\mu>$ denotes the Wannier state associated with site i in band μ (μ=c or v). The diagonal element E_i^μ is the random site energy, and V_{ij}^μ represents the regular transfer energy. Within the approximation of a constant dipole matrix element, the imaginary part of the dielectric function $\varepsilon_2(\hbar\omega)$ is formulated as [11]

$$\varepsilon_2(\hbar\omega) \propto \iint dE_1 dE_2 \, \delta(E_1 - E_2 - \hbar\omega)$$

$$\times \ (K(E_1^+, E_2^+) - K(E_1^+, E_2^-) - K(E_1^-, E_2^+) + K(E_1^-, E_2^-)), \tag{6}$$

where $E^{\pm} = E \pm i0$, and

$$K(z_1, z_2) \equiv \Sigma_{ij} <<G_{ij}^{\ c}(z_1) G_{ji}^{\ v}(z_2)>>. \tag{7}$$

$G_{ij}^{\ \mu}(z)$ denotes the single-particle Green function in the Wannier representation;

$$G_{ij}^{\ \mu}(z) = <i\mu|(z-H^{\mu})^{-1}|j\mu>, \tag{8}$$

and $<<....>>$ represents the ensemble average over the distribution of the site energy E_i^{μ}.

We model the QW system with infinite potential barriers by a simple cubic lattice truncated in <001> direction. Namely, the site position is specified by $R_i = (ap, aq, ar)$, where a is the lattice constant, and $-\infty < p, q < +\infty$, $r = 1, 2, ..., N$, so that aN corresponds to the well-layer width L_W. The electron state with the subband index n and the well-plane parallel k vector is thereby represented in terms of the Wannier basis $|pqr:\mu>$;

$$|nk:\mu> = c\Sigma_{pqr} \sin(n\pi r/(1+N)) \exp(ikR_{pq}) |pqr:\mu>, \tag{9}$$

where c is a normalization constant. The regular transfer energy part of the Hamiltonian (5) is found to be diagonal in the nk-representation, leading to the eigenenergy

$$E_{nk}^{\ \mu} = 2V^{\mu}(\cos k_x a + \cos k_y a + \cos(n\pi/(N+1))). \tag{10}$$

If the statistical correlation between the site-diagonal terms $E_i^{\ c}$ and $E_i^{\ v}$ is ignored, the ensemble average of the product of the Green functions, eqn. (7), becomes identical with the product of the averaged Green functions $<<G_{ij}^{\ c}(z_1)>> <<G_{ji}^{\ v}(z_2)>>$. In the calculation, the averaged Green function is derived within the framework of the Coherent

Potential Approximation (CPA) [11,12]. Since the CPA is a single-site approximation, the nk-representation of the Green function defines a self energy $\Sigma^\mu(E)$ which is independent on n and k, and we have

$$K(z_1, z_2) = \Sigma_{nk} G_{nk}^c(z_1) G_{nk}^v(z_2), \tag{11}$$

where

$$G_{nk}^\mu(z) \equiv <nk:\mu|<<(z-H^\mu)^{-1}>>|nk:\mu>$$
$$\equiv (z-\Sigma^\mu(E)-E_{nk})^{-1}. \tag{12}$$

In the calculation, we deal with the case of Gaussian site disorder, in which the standard deviation of the distribution is described by W for both the conduction and valence bands. The degree of disorder is represented by the ratio of W to the full band width B which is 2x6xV in the case of N→∞. Here, for the sake of simplicity, the same band width is assumed for both the conduction and valence bands.

Figure 6 displays how the absorption spectrum of the QW structure near the direct edge varies with the degree of disorder W/B. The spectra are given by αħω assuming a constant refractive index, and the direct gap of the unperturbed bulk (N→∞) system is set at 2eV. It is found that the spectral shape changes from a staircase form being characteristic to crystalline QWs to a structureless form as the degree of disorder increases. For W/B≥0.06, the Tauc plot forms a straight line in an appreciable range of photon energy, giving the Tauc gap, which is indicated by arrows in the figure.

The differential absorption spectrum corresponding to the PTM spectrum can be easily calculated by differentiating $\varepsilon_2(\hbar\omega)$ by ħω, and the result is shown in Figure 7. Again the vertical scale is given by the differential absorption Δα multiplied with ħω. As found, the spectrum exhibits well-resolved peaks at each subband transition edge when the degree of disorder is small. As W/B gets larger and

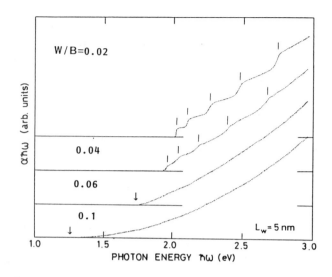

FIG. 6. Absorption spectra calculated for QW with the well width L_W=50Å. The parameter W/B represents the degree of disorder introduced to the site energy, and arrows indicate Tauc gap.

approaches 0.06, then the spectrum appears to reveal a broadened staircase form. Finally, structures almost disappears from the spectrum when W/B becomes larger than about 0.1. The overall behavior of the absorption spectrum calculated for W/B=0.06 agrees well with that experimentally observed in a-Si:H/a-SiC:H QW structures; features associated with subband transitions are hardly found out in the absorption spectrum, while they clearly dominate the spectral shape as a staircase form in the differential absorption spectrum. The result of self-consistent calculation seems to confirm the plausibility of the interpretation of the experimental results on the basis of the simpler model described in the previous section. Another conclusion reached here that the degree of disorder W/B may be around several percents in the realistic a-Si:H coincides with that suggested by Yonezawa [12] from the aspect of the absorption tail width.

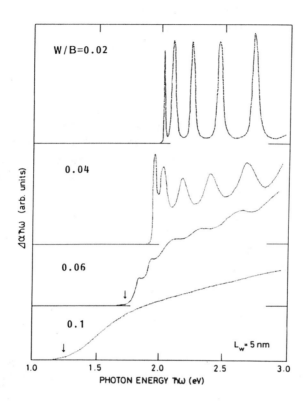

FIG. 7. Differential absorption spectra of QW with the well width L_w=50Å. Note that quantitative comparison with the experimental results in Fig. 5 is not possible, since the calculation relies on a model density of states associated with isolated QW consisting of a simple cubic lattice.

The degree of disorder on the order of W/B≅0.06 implies electron coherent length in the range between 10 and 20Å near the band edge.[13] This extent is likely to be consistent with suggested electron microscopic mobility on the order of $10 cm^2 V/s$ [14], although the conventional picture of carrier transport on the basis of the random phase approximation [15] can not be immediately applied. If the evaluation of the coherent length is correct, then a considerable alternation should be required on the interpretation of the resonant tunneling phenomenon observed in a-Si:H double-barrier QW structures since the wave vector parallel to the well-plane may not be conserved over the typical dimension of the heterostructures.

5. CONCLUSION

Differential absorption spectra of the a-Si:H QW structures have been investigated with varying the layer thickness. When the a-Si:H layer thickness decreases to less than 50A, the spectrum appears to exhibit a staircase behavior, indicating that the valence and conduction bands split into 2D-subbands. The self-consistent calculation for the optical absorption has been carried out, confirming the validity of the conclusion reached by the simple consideration starting with a completely broken k-selection rule. It is also demonstrated that an extreme detection sensitivity for unresolved optical structures achieved in differential spectroscopy provides a powerful way to accurately determine the subband transitions.

The experimental data for the energy shift of subband transitions are in quantitative agreement with those expected from the square QW model. Electron effective mass is suggested to be $0.3m_0$. It is of interest to note that the directional averaged effective mass takes a value close to $0.3m_0$ in crystalline Si, SiC (3C and 6H), C (diamond), and SiO_2. The implication from this apparent coincidence might be important for understanding the electronic structure of a-Si:H and its alloy materials.

REFERENCES

1. For example, Abeles, B. and Tiedje, T., Semiconductors and Semimetals (Academic, New York, 1984), Vol. 21, Pt. C, p. 407.
2. Miyazaki, S., Ihara, Y. and Hirose, M., Phys. Rev. Lett. $\underline{59}$, 125 (1987).
3. Cardona, M., Modulation Spectroscopy (Academic, New York, 1969).

4. For example, Frova, A. and Selloni, A., Tetrahedrally-Bonded Amorphous Semiconductors (Plenum, New York and London, 1985), p. 271.
5. Dutta, N. K., J. Appl. Phys. 53, 7211 (1982).
6. Hattori, K., Mori, T., Okamoto, H. and Hamakawa, Y., Phys. Rev. Lett. 60, 825 (1988).
7. Hattroi, K., Mori, T., Okamoto, H. and Hamakawa, Y., Advances in Amorphous Semiconductors I, Amorphous Silicon and Related Materials, edited by Fritzsche, H. (World Scientific, 1988), p. 957.
8. Pfost, D., Hsiang-na Liu, Vardeny, Z. and Tauc, J., Phys. Rev. B30, 1083 (1984).
9. Hattori, K., Mori, T., Okamoto, H. and Hamakawa, Y., Appl. Phys. Lett. 51, 1259 (1987).
10. Hattori, K., Mori, T., Okamoto, H. and Hamakawa, Y., Appl. Phys. Lett. 53, 2170 (1988).
11. Abe, S. and Toyozawa, Y.,J. Phys. Society of Japan 50, 2185 (1981).
12. Yonezawa, F. and Sato, F, 57, 1797 (1988).
13. Economou, E. N., Soukoulis, C. M., Cohen, M. H. and Zdetsis, A. D., Phys. Rev. B31, 6172 (1985).
14. Economou, E.N., Physics and Applications of Amorphous Semiconductors, edited by Demichelis, F. (World Scientific, 1987), p. 3.
15. Hindley, N. K., J. Non-cryst. Solids, 5, 17 (1970).

Growth of Polycrystalline Diamond Films by Filament Assisted CVD of Hydrocarbons

T.D. Moustakas
Boston University, College of Engineering
Boston, MA 02215

R.G. Buckley
Physics and Engineering Laboratory
DSIR, PO Box 31313, Lower Hutt, New Zealand

ABSTRACT

Polycrystalline diamond films were deposited by the method of W-filament-assisted CVD of mixtures of methane and hydrogen. It was found that the W-filament undergoes carburization and converts into $\alpha - W_2C$. This process affects the nucleation of the diamond films. The structure of the films was investigated by Raman Spectroscopy and Electron Microscopy. Both of these studies reveal the existence of large number of point and line defects, whose density increases as the growth pressure decreases, and eventually the films undergo a transformation to a disorder phase made of mixtures of disorder diamond and graphite.

1. Introduction

The growth of diamond thin films at low pressure and temperature, utilizing vapor phase methods, is both an interesting problem in the science of crystal growth and a process with significant commercial potential as an alternative to growing this material in the form of powders by the high pressure and temperature method [1]. One can envision that thin films of diamond will find applications related to high hardness, chemical inertness, high thermal conductivity, and to unique semiconducting properties.

Early studies on the synthesis of diamond thin films [2,4] suggested that pure thermal decomposition of hydrocarbons favors the deposition of graphitic carbon, which is the thermodynamically stable phase under low pressure conditions. In the past few years it was realized [5,6] that decomposition of hydrocarbons in the presence of atomic hydrogen can lead to pure diamond films. This process combines deposition of carbon in its graphitic and diamond forms and etching of carbon by atomic hydrogen. While the deposition process favors the formation of graphite, the H-etching process favors the gasification of graphite. An appropriate balance between these processes could lead to the formation of pure diamond.

The deposition methods proposed are various forms of assisted chemical vapor deposition of mixtures of hydrogen with hydrocarbons. Such method are the Chemical Transport Reaction [6], RF plasma CVD [7], microwave-plasma CVD [8], laser assisted CVD [9], and filament assisted CVD [10]. All of these processes were optimized to produce atomic hydrogen from molecular hydrogen and to activate the hydrocarbon species into the appropriate carbon precursors.

In the filament assisted CVD process the tungsten filament was originally introduced to dissociate molecular hydrogen and form atomic hydrogen. The catalytic dissociation of molecular hydrogen on a hot tungsten surface was first discovered by Langmuir [11] and studied in more detail recently by a number of other workers [12]. However, besides this role, it was discovered the past two years [13-15] that during diamond deposition by the filament assisted CVD process the tungsten filament also undergoes carburization, a process which affects the structure and microstructure of the filament and influences the nucleation process of the diamond films.

In this paper an overview of our own studies of the W-filament-assisted CVD process for the growth of diamond thin films is presented. More specifically, we discuss the process of thin film growth as well as the structure of the films.

2. Experimental Methods

The experimental apparatus for the synthesis of the diamond films is shown in Figure 1. The vacuum chamber consists of a stainless steel vessel 30 in. in diameter and 12 in. high. Prior to the introduction of the reaction gases this chamber was pumped by a combination of turbomolecular and mechanical pumps to a total leak rate of 2×10^{-5} Torr·l·s^{-1}. The methane and hydrogen gases were either supplied from different tanks or were pre-mixed in one tank. During the introduction of the reaction gases the turbomolecular pump was by-passed and the desired total pressure was attained by throttling the mechanical pump.

The substrates were placed onto a molybdenum block 1.5 in^2, which was heated radiatively. The substrate temperature was determined with a Chromel-Alumel thermocouple embedded in the molybdenum block. The substrate holder was either electrically grounded or positively biased. The substrates used in this study were either polycrystalline Mo−plates or $Si(100)$ single crystal plates. Some of the substrates were blasted with SiC powder (25μ particles) to increase the nucleation density.

The tungsten filament, 0.5mm in diameter, consists of four straight segments covering uniformly the substrate holder at about 1.0cm above the substrates. The filament was heated with a DC power supply and its temperature was monitored with an optical pyrometer.

The range of depositon parameters investigated were substrate temperature (750^0 to 950^0C), filament temperature (1600 to 1800^0C, not corrected for filament emissivity), methane concentration (0.5 to 2.0 Vol%) and total gas pressure (1 to 100 Torr).

The tungsten filaments as well as the deposited films were characterized by X-ray diffraction, scanning electron microscopy (SEM), transmission electron microscopy (TEM) and Raman spectroscopy. Details of the TEM and Raman Spectroscopy were published elsewhere [16,17].

3. Experimental Results and Discussion

Figure 2 shows the cross-sectional structure of a tungsten filament before and after being heated to 1700^{0}C in a mixture of 0.5 vol% CH$_4$ in hydrogen at a total pressure of 30 Torr for 5 hours. The microstructure of the filament before the treatment is columnar with the columns oriented along the length of the wire. After the treatment the wire develops large grains oriented towards the center of the wire.

X-ray diffraction studies of powders of the treated filament, shown in Figure 3, reveal that the filament has been converted from W to alpha-W_2C. The carburization of the filament accounts for the change of the filament microstructure, since the absorption of carbon proceeds from the perimeter towards the center of the wire. Additionally, the volume expansion, due to carbon incorporation, leads to the observed cracks. These findings, are in agreements with those by Suzuki et. al. [13].

The kinetics of this carburization process can be studied by monitoring the resistance of the filament as a function of time. Such studies were reported elsewhere [15] and indicate that the filament becomes fully carburized in approximately 10 minutes. The length of this time varies with the relative concentration of methane in hydrogen and the temperature of the filament. These findings, in agreement with those of Suzuki et. al. [13], suggest that the carburization of the filament results in consumption of the carbon from the CH_4 and thus an incubation time is needed for the nucleation of the diamond films.

The process of the filament carburization also affects its electron emission properties. As shown elsewhere [15], this result affects the nucleation of the diamond films, since the rate of nucleation depends on the flux of the electrons impinging upon the substrate.

When the substrate holder was electrically grounded continuous films were formed only on roughened substrates and isolated crystals were formed on polished substrates. Films grown on positively biased substrates were found to be continuous, independent of the substrate treatment. Figure 4 shows isolated crystals on a polished silicon substrate grown on an electrically grounded substrate holder. Figure 5 shows a continuous film on a polished substrate grown on a positively biased substrate holder. It is obvious from these results that biasing the substrate positively increases the electron bombardment of the substrate, which leads to higher nucleation rate for diamond crystals. These findings agree with those of Sawabe and Inuzuka [18]. Figure 6 shows a continuous films grown on to a roughened substrate, under the same conditions as the sample discussed in Figure 5. It is apparent that the microstructures revealed in Figures 5 and 6 are different, underlying differences in the nucleation and growth mechanisms.

A larger number of these films were investigated by Raman Spectroscopy. The details of these studies were reported elsewhere [17]. In here we present only a brief summary of these results. The films for this study were produced under the following conditions: The filament assisted CVD process took place in a gas mixture of CH_4 (2%) and H_2 (98%) with total gas pressures varying from 5 torr to 100 torr. The tungsten filament was operated at 1800°C (not corrected for filament emissivity) 10 mm above the substrate, which was positively biased relative to the filament. The emission current was held constant at 13 mA/cm^2 by adjusting the substrate bias voltage. All the results reported here are for films prepared on Mo substrated blasted with SiC powder (25 μ particles). Films prepared on Si substrates did not display significantly different Raman spectra.

Displayed in Figure 7 are Raman spectra for a series of films prepared at a range of total gas pressures with all the other preparation parameters held constant, in particular, the CH_4 to H_2 ratio. The spectra are taken from the same area of the samples relative to the filament. Overall the spectra demonstrate the multiphase nature of these films and the strong dependence of these phases on total gas pressure. The existence of a relatively sharp feature near 1330 cm^{-1} demonstrates the presence of diamond in all the CVD films. This is the triply degenerate phonon of F_2g symmetry normally reported to be near 1332 cm^{-1} in good quality single crystal diamond [19,20]. These spectra were collected at a relatively low resolution ($\sim$ 8 cm^{-1}) so that the line suffers from instrument broadening. The other features in the spectra are characteristic of non-diamond forms of carbon and are discussed below. For the film prepared at 5 torr the diamond peak is nearly obscured indicating that diamond contributes only a minor fraction to this film.

The spectra of films prepared at pressures less than 30 torr display a very broad feature centered about 1550 cm^{-1}. A second broad peak can also be observed near 1360 cm^{-1} that grows in relative intensity and shifts to lower frequency with decreasing pressure. Both these features have been reported previously in diamond films [21,22] and indicate the presence of a disordered carbon component in the films [23-25]. At higher preparation pressures ($>$ 30 torr) the high frequency non-diamond peak in the spectra sharpens noticeably, shifts to a higher frequency and splits into two peaks at 1578 cm^{-1} and 1546 cm^{-1}. The 1578 cm^{-1} peak can be assigned to the E_2g mode of crystalline graphite [26,27], however, the peak at 1546 cm^{-1} cannot be easily assigned, but its magnitude is very sensitive to the preparation gas pressure, and is strongest in the film prepared at the immediate pressure of 60 torr.

Figure 8 displays the diamond line for the same series of films, but at a higher resolution and compares them with the measured line for a natural diamond crystal. The figure demonstrates firstly, that the line widths for these films are significantly greater than for natural diamond and secondly, there is a strong dependence of line width on total gas pressure. For instance, changing the preparation conditions from low pressure ($\sim$ 20 torr) to high pressure ($>$ 50 torr) results in about a factor of two reduction in line width.

Figure 8 also illustrates that except for the film prepared at 15 torr the position of the line is nearly independent of pressure and that it lies at a lower frequency than the natural diamond line. The diamond line for the 15 torr sample lies at a higher frequency than the natural diamond line indicating that this film may be under stress.

Figure 9 displays two SEM micrographs typical of the investigated films. Except for the film prepared at 5 torr all films are dense and display well faceted crystals. The film prepared at 5 torr displays a ball-like morphology and the Raman spectra of this film is dominated by structures typical of amourphous carbon.

The spectra of Figures 7 and 8 display a number of features that have been observed before and are readily assigned as was done above. However, the peak at 1546^{-1} can only be tentatively assigned. During a Raman study of ^{11}B ion implanted graphite, Elman et al [28] observed that with increasing ion fluence at high fluences, a continuous broadening and down shifting of the 1580 cm^{-1} graphite line resulted from an increasing disorder in their sample. The present data allow as for a similar interpretation. At the highest pressure (100 torr) the presence of a relatively intense peak at 1579 cm^{-1} indicates a large crystalline graphitic component while the well developed feature at $\sim$ 1360 cm^{-1} implies the crystals of this component are small [26]. This film also displays a shoulder at 1546 cm^{-1}. However, as the gas pressure decreases the 1546 cm^{-1} peak increases rapidly in intensity and at the lowest pressures both the 1579 cm^{-1} and 1546 cm^{-1} peaks are replaced by the commonly observed single broad peak in disordered carbon at $\sim$ 1550 cm^{-1}. The 1360 cm^{-1} peak also broadens as the pressure decreases. Both these features dominate the spectra at the lowest pressure of 5 torr and indicate a large amorphous component [23,24]. We interpret the 1546 cm^{-1} peak as implying the presence of crystalline graphite with a large number of defects, but with decreasing total gas pressure the defect density grows until the graphite phase becomes completely disordered. The presence of peaks at 1580 cm^{-1} and 1546 cm^{-1} in the films prepared at high pressure indicates the presence of two distinct crystalline graphite components that differ in the degree of disorder in each.

Accompanying the increasing disorder with decreasing pressure in the graphite component there is also evidence for increasing disorder in the diamond component. As noted above the width of the diamond line is always larger than that of natural diamond and further, the line width increases dramatically at low gas pressure where it is nearly four times the natural diamond line width. A line broadening of this magnitude indicates a density of defects such that phonon lifetimes are significantly reduced and with the implication that the defect density increases with decreasing pressure. As the diamond line position is approximately independent of gas pressure and close to the frequency expected for natural diamond, it would appear that the disorder is not affecting the average bond strength of the crystallites that make up the film.

The most notable aspect of this study is that the filament assisted CVD technique of diamond film production results in a variety of carbon types, even though the ratio of CH_4 to H_2 was constant. The relative contribution to the film of any single carbon type depends strongly on the total gas pressure such that the more disordered types of carbon form a larger component for films prepared at the lowest pressures. This process of increasing disorder with decreasing pressure results, at the lowest pressure of 5 torr, in a film that appears to be almost completely disordered carbon. The film preparation conditions at relatively low temperature and pressure are such that diamond is not thermodynamically favored and this results in a large density of defects in the crystalline diamond component. During film growth there is competition between diamond and graphite growth, and except at the lowest pressure (5 torr) diamond predominates due to the higher etching rate of graphite by atomic hydrogen. The flux of atomic hydrogen reaching the substrate increases with decreasing pressure due the longer mean free path which may account for the observed smaller graphitic component and the accompanying increase in the amorphous carbon content. This is illustrated by the Raman results where the crystallinity of both the graphite and diamond phases decrease with decreasing total gas pressure and ultimately produce an almost completely disordered carbon film at 5 torr. In particular, the quality and relative contribution of diamond to the film depends on the total gas pressure.

The existence of a large number of defects in these polycrystalline diamond films was also revealed in electron diffraction and bright field TEM studies [16]. These studies reveal regions with large number of stacking faults and twins.

4. Conclusions

In the filament assisted CVD of mixture of hydrocarbons with hydrogen for the formation of diamond films the tungsten filament undergoes carburization and converts into alfa-W_2C. During this stage a significant fraction of the CH_4 is consumed by the filament, and film growth is slow.

The Raman spectroscopy of the deposited films probes the quality of diamond films, in particular the crystallinity of the diamond phase. It is observed that as the growth pressure is reduced the density of lattice defects in both the diamond and graphite components are increased such that at the lowest pressures (5 torr) a completely disordered carbon film results. The high flux of neutral and charged particles at low pressures is likely to result in the increase defect density in the diamond component.

References

1. F.P. Bundy, H.T. Hall, H.M. Strong and R.H. Wentorf Jr., Nature, 176, 51 (1955)

2. J.C. Angus, H.A. Will and W.J. Stanko, J. Appl. Physics, 39, 2915 (1968)

3. D.J. Poferl, N.C. Gardner and J.C. Angus, J. Appl. Physics, 44, 1428 (1973)

4. B.V. Deryaguin, D.V. Fedroseev, V.M. Lykuanovich, B.V. Spitsyn, V.A. Ryanov and A.V. Laurentyev, J. Crystal Growth, 2, 380 (1968)

5. B.V. Deryaguin, B.V. Spitsyn, L.L. Bouilov, A.A. Klochkov, A.E. Gorodestskii and A.V. Smolyanov, Sov. Phys. Dokl., 21, 676 (1976)

6. B.V. Spitsyn, L.L. Bouilov and B.V. Derjaguin, Journal of Crystal Growth, 52, 219 (1981)

7. S. Matsumoto, J. Matls. Sci. Lett., 4, 600 (1985)

8. M. Kamo, Y. Sato, S. Matsumoto and N. Setaka, J. Cryst. Growth, 62, 642 (1983)

9. K. Kitihama, K. Hirata, H. Nakamatsu and S. Kawai, Appl. Phys. Lett., 49, 634 (1986)

10. S. Matsumoto, Y. Sato, M. Kamo and N. Setaka, Jap. J. Appl. Phys., 21, L183 (1982)

11. L. Langmuir, J. Am. Chem Soc., 34, 1310 (1912)

12. T.W. Hickmott, J. of Chemical Phys., 32, 810 (1960)

13. H. Suzuki, H. Matsubara and N. Horie, Futai Oyaobi Funtai Yakin, 33, 281 (1986)

14. T.D. Moustakas, J.P. Dismukes, Ling Ye, K.R. Walton and J.T. Tiedie, Proc. of 10th International Conf. on CVD, Honolulu 1987 (The Electrochemical Society Inc. 1987) pp. 1164-1173

15. T.D. Moustakas, Solid State Ionics (1989)

16. Mark M. Disko and T.D. Moustakas, Proc. of MRS Meeting (Boston, 1988)

17. R.G. Buckley, T.D. Moustakas, Ling Ye and J. Varon, J. Appl. Phys., (in press)

18. A. Sawabe and T. Inuzuka, Appl. Phys. Lett., 46, 146 (1985)

19. R.S. Krishnan, *Proc. Indian Acad. Sci. Sec.* A24, 45 (1946).

20. S.A. Solin and A.K. Ramdas, *Phys. Rev.* B1, 1687 (1970).

21. R.C. DeVries, *Ann. Rev. Mater. Sci.* 17, 161 (1987).

22. S. Matsumoto, M. Hino, and T. Kobayashi, *Appl. Phys. Lett.* 51, 737 (1987).

23. N. Wada, P.J. Gaczi, and S.A. Solin, *J. Non-cryst. Solids* 35 & 36, 543 (1980).

24. R.O. Dillon and J.A. Woollam, *Phys. Rev.* B29, 3482 (1984).

25. M. Ramsteiner and J. Wagner, *Appl. Phys. Lett.* 51, 1355 (1987).

26. F. Tuinstra and J.L. Koenig, *J. Chem. Phys.* 53, 1126 (1970).

27. R.J. Nemanich and S.A. Solin, *Phys. Rev.* B20, 392 (1979).

28. B.S. Elman, M. Shayegan, M.S. Dresselhaus, H. Mazurek, and G. Dresselhaus, *Phys. Rev.* B25, 4142 (1982).

Figure Captions

Figure 1. Schematic diagram of the deposition system.

Figure 2. Cross-sectional structure of a tungsten filament before and after treatment in a $CH_4 + H_2$ misxture as described in the text.

Figure 3. X-ray poweder diffraction pattern of the filament described in Figure 2.

Figure 4. Isolated diamond crystals on a polished silicon (100) substrate, grown on electrically grounded substrate holder.

Figure 5. Continuous diamond film on a polished silicon (100) substrate, grown on a positively biased substrate holder.

Figure 6. Continuous diamond film on a roughened silicon (100) substrate, grown on a positively biased substrate holder.

Figure 7. A plot of the Raman spectra for filament assisted CVD diamond films prepared at the same CH_4 to H_2 ratio but with changing total gas pressure. The line at ~ 1330 cm^{-1} is due to that of diamond while the line at ~ 1580cm^{-1} indicates the presence of graphite. The broad features result from disordered carbon, a component that increases with decreasing total gas pressure. [after reference 17]

Figure 8. A plot that compares the position and line width of the diamond line for filament assisted CVD films with that of natural diamond (the narrow line). The figure shows that the diamond line width for the films is wider than for natural diamond and is shifted in frequency. The insert shows the dependence of the line width on total gas pressure. [after reference 17]

Figure 9. Typical SEM micrographs of the diamond films discussed in the text: (a) prepared at 5 torr, (b) prepared at 15 torr. Film prepared at higher pressures show similar microstructures as those shown in Figure (b).

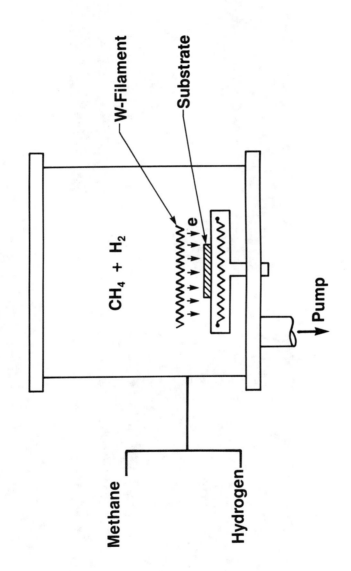

Figure 1.

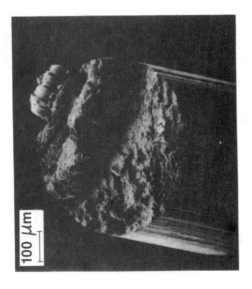
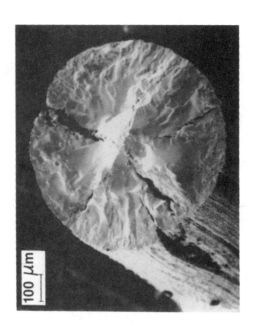

Figure 2.

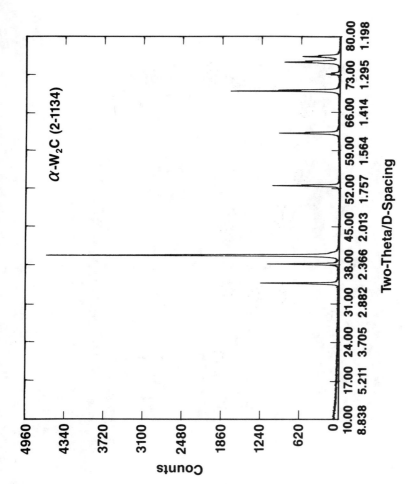

Figure 3.

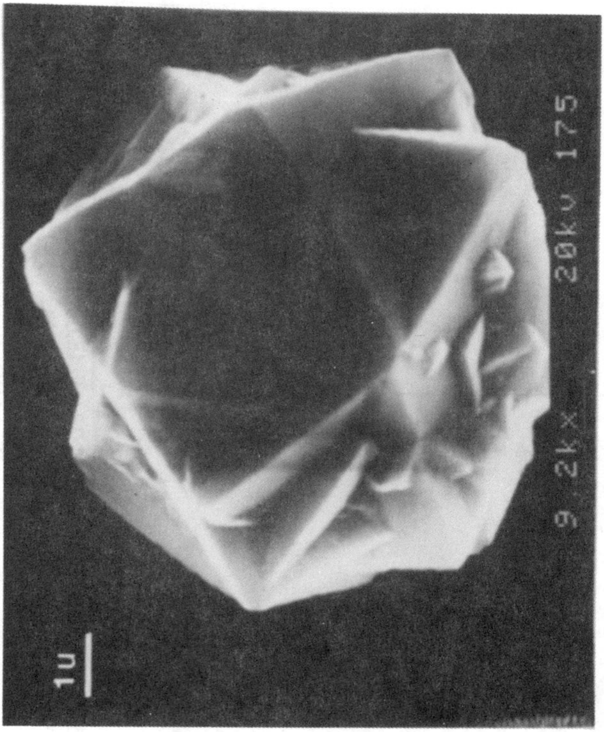

Figure 4.

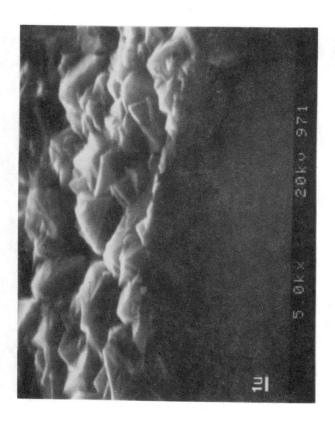

Figure 5.

Figure 6.

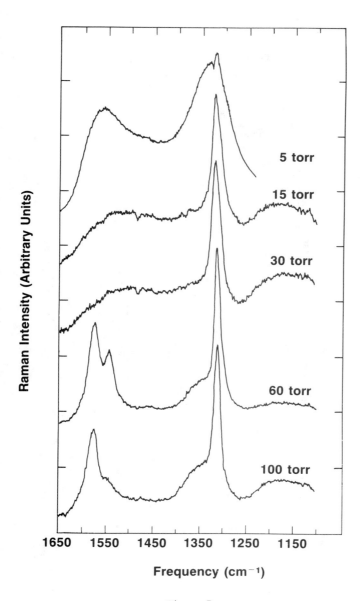

Figure 7.

126

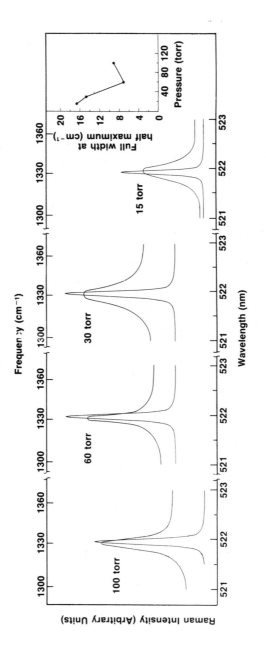

Figure 8.

Figure 9.

GROUP IV ELEMENTS IN AMORPHOUS SEMICONDUCTOR ALLOYS

F. DEMICHELIS and A. TAGLIAFERRO

Dip. Fisica - Politecnico Torino - Torino (ITALY)

INTRODUCTION

In the past few years amorphous silicon based alloys were the subject of extensive research as semiconducting materials of both technological and basic scientific importance.

a-SiGe:H and a-SiSn:H alloys were investigated as materials having narrower band gap and a-SiC:H as a material having wider band gap than a-Si:H. Among group IV elements, carbon has received relatively little attention compared with Si and Ge. Only in recent times, carbon based alloys such as a-CSn:H, a-CSiSn:H and a-CSiGe:H have been realized and characterized [1,5].

The presence of carbon in a-SiSn:H and a-SiGe:H gives rise to the peculiar characteristcs of such ternary carbon based alloys. While silicon and germanium are fully stable only under fourfold coordination, carbon can form molecular bonds in any of the three sp^3, sp^2 and sp^1 configuration, so that different structures, either diamond (pure sp^3), graphite (pure sp^2) or amorphous carbon (mixture of sp^3 and sp^2), exist.

In this paper we report some results of studies performed on amorphous ternary alloys: a-CSiSn:H and a-CSiGe:H. We present dark conductivity vs temperature data for a series of alloys of each type as a function of carbon content and the changes in the average coordination number as the carbon content is increased. The results show how much the presence of carbon affects the physical properties of a-CSiSn:H and a-CSiGe:H alloys.

PREPARATION AND CHARACTERIZATION TECHNIQUES

Samples of a-CSiSn:H and a-CSiGe:H have been deposited by Sputter Assisted Plasma Chemical Vapour Deposition (SAPCVD). The magnetron cathodes have been Sn (purity 99.9995%) or Ge (purity 99.999%) targets. A mixture of SiH_4, CH_4 and Ar has been introduced into the chamber, controlling the flow of each gas separately. The Ar and SiH_4 flows have been kept constant while the CH_4 flow was varied in the range 0-10 sccm. The R.F. (13.56 MHz) power, the pressure, the deposition temperature (120°C for a-CSiSn:H and 100°C for a-CSiGe:H) have been chosen on the ground of previous results[2,3,5].

The resistances have been measured as a function of temperature in a vacuum chamber with a Hewlett Packard high resistance meter 4329A. A dc voltage in the range 10 to 500 V in direct and reverse bias has been applied on aluminium electrodes (8 mm long and with a spacing between them of 1 mm) pre-evaporated in a coplanar arrangement onto Corning 7059 glass substrate.

Elemental composition has been determined by means of RBS and ERDA analysis[3].

AVERAGE COORDINATION NUMBER

The average coordination number is of particular interest because of recent theories[6,7] on random covalent networks. Once the average coordination number is defined as

$$m = \frac{\sum r\, n_r}{N}$$

where N is the total number of atoms in the network formed by n_r atoms having r bonds, these theories indicate the optimum average coordination number as given by $m_{opt} = 2.45$.

In this optimal condition the stabilizing influence of bonding balances the destabilizing influence of strain energy. Amorphous solids with average coordination numbers much greater than the optimal value would be highly "over-constrained" and would act to relieve strain by forming islands surrounded by intrisic dangling bonds.

The average coordination number in a-Si or a-Ge can be reduced by the addition of hydrogen (n=1), but for an hydrogen content of 0.30 it is reduced to 3.1 only.

In amorphous carbon the average coordination number can be reduced by the addition of hydrogen or by the presence of significant amounts of sp^2 bonding (m=3). For polymeric hydrogenated amorphous carbon films the proportionality $sp^3:sp^2:sp^1$ = 53:45:2 holds[8]. In such a case, average coordination number reduces to 2.78 for a 0.30 content of hydrogen.

We have applied the Phillips-Thorpe model in order to calculate the average coordination number of our a-CSiGe:H and a-CSiSn:H alloys using the data on the composition of the samples[3,5]. Results are shown in Fig. 1, where the average coordination number is reported as a function of the carbon content. It can be observed that the average coordination numbers of the two different alloys are differently correlated to the carbon content. This is a clear sign of the intrinsic difference of these two alloys. In fact, as a general rule, a larger amount of hydrogen is embedded in a-CSiGe:H than in a-CSiSn:H[3,5].

It is clear that, increasing carbon content, the average coordination number is moved towards the Phillips-Thorpe maximum stability value.

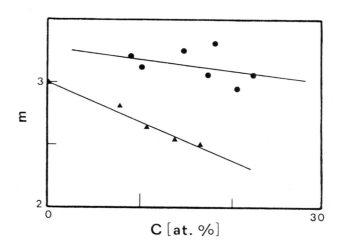

FIG. 1 Average coordination number for a-CSiSn:H (●) and a-CSiGe:H (▲)

TEMPERATURE DEPENDENCE OF CONDUCTIVITY

Fig. 2 shows the dark conductivity as a function of inverse temperature for two typical samples of a-CSiGe:H alloys having a different carbon content. The samples are labeled SiGe0 and SiGe17, where number indicates the atomic percentage of carbon$^{(5)}$: we can observe a remarkable difference in the behaviour of the two samples, with and without carbon content. In the SiGe17 sample, before heating, the conductivity is high and remains so until about 110°C, when it starts to decrease; above a defined temperature (which value depends upon carbon content$^{(5)}$) the conductivity increases once again, becoming independent of thermal hystory, as confirmed by the trend in the cooling process and in successive cycles. a-CSiSn:H samples show quite similar behaviours, when Sn content is low (below 2 at%).

By comparison of a-CSiGe:H curves with a-SiGe:H curve, we deduce that the presence of carbon modifies the electrical properties of the binary alloy a-SiGe:H.

As shown in Fig. 2, the curves above and below the hump can be described by an Arrhenius relationship

$$\sigma_d = \sigma_0 \exp(-E_a/kT)$$

with two different sets of preexponential factors (σ_0) and activation energies (Ea) values.

At low temperature (below deposition temperature) activation energies of 0.66 and 0.31 eV and preexponential factors of $2.8 \cdot 10^2$ $\Omega^{-1}cm^{-1}$ and $1.4 \cdot 10^{-3}$ $\Omega^{-1}cm^{-1}$ for SiGe0 and SiGe17 respectively have been obtained. At higher temperatures Arrhenius behaviour is restored, with $E_a = 0.95$ and $\sigma_0 = 5.9 \cdot 10^5$ for SiGe0 and $E_a = 0.96$ and $\sigma_0 = 9.5 \cdot 10^2$ for SiGe17 respectively.

From these results and from the results obtained for other films of different composition $^{(5)}$, it is clear that the hump in conductivity and the carbon content are closely related.

Let us now try an interpretation of the experimental data. In the

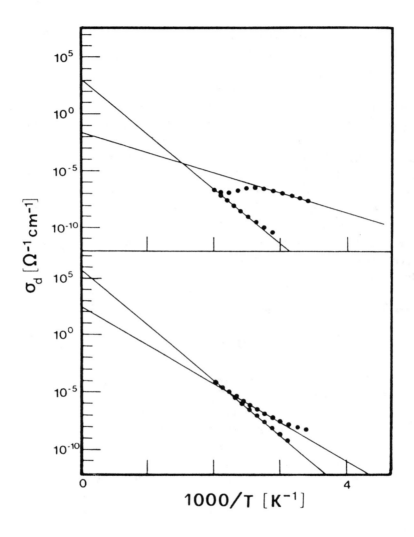

FIG. 2
Dark conductivity behaviour for a-CSiGe:H with (upper curve) and without carbon (lower curve)

heating process, when the temperature becomes higher than deposition temperature, the free hydrogen embedded in the alloy tends mostly to bond with carbon, predominantly in CH_2 form, rather than with Si or Ge (or Sn). I.R. measurements performed on the samples before and after annealing [9] confirm this change in structure. The increase of sp^3CH_2 configuration favours sp^3 over sp^2 bonding and a consequent reduction of π electrons and in turn of conductivity. Above a certain temperature (related to the carbon content), the activated conductivity behaviour is restored and its characteristic values are similar to that of SiGeCO, a material which has no carbon. The temperature at which Arrhenius behaviour is restored corresponds to the threshold temperature at which the percolative conductivity of carbon drops below the conductivity of the host material a-SiGe:H.

An interesting point is that a-CSiGe:H samples show photoconductivity after annealing at 250°C. For samples with carbon content of 8 at% the σ_{ph}/σ_d ratio is $5 \cdot 10^2$ while for samples a-SiGe:H is $\sigma_{ph}/\sigma_d = 1.3 \cdot 10^3$, both under AM2 illumination

Since activation energy is almost independent from composition and preexponential factor is not, the Meyer-Neldel rule

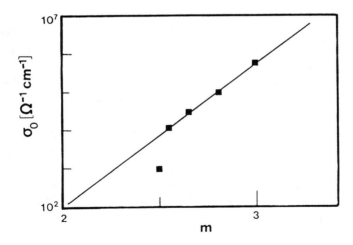

FIG. 3   Influence of average coordination number upon dark conductivity pre-exponential factor in a-CSiGe:H

$$\sigma_o = \sigma_{oo} \exp(E_a/kT_o)$$

is not verified for these materials, if σ_{oo} and T_o are assumed to be the same for all samples. However, fig.3 clearly shows that preexponential factor is not an independent parameter: its dependence upon the average coordination number is striking. The point representing SiGe17 does not fit the exponential relationship: our explanation is that the maximum annealing temperature is not sufficient to completely equilibrate the material, so that σ_o, which is increasing (compare the two values in the low and in the high temperature regions given a few lines above), has not yet reached its final value. The behaviour observed in fig.3 is consistent with the assumption that conduction after annealing is mainly due to the silicon-germanium hydrogenated matrix.

CONCLUSION

The temperature dependence of dark conductivity and the average coordination number in a-CSiGe:H and a-CSiSn:H random networks have been studied and analyzed. We have shown that:
. the ternary based carbon alloys have an average coordination number which tends towards the optimal value predicted for amorphous covalent networks, as carbon content increases.
. the temperature dependence of dark conductivity is strongly affected by the carbon content and more precisely by the presence of sp^2 sites in the carbon bonds
. the relationship between dark conductivity pre-exponential factor and the average coordination number indicates that the a-SiGe:H matrix is mainly responsible for post annealing conduction

From the above results we can deduce that the proportion $sp^3:sp^2$ sites in carbon is responsible of the unusual properties of the as deposited carbon based alloys.

REFERENCES

1) F. Demichelis, G. Kaniadakis, P. Mpawenayo, M.A. Perino, A. Tagliaferro, E. Tresso, P. Rava, G. Della Mea, M. Vallino
   Thin Solid Films 150, 189 (1987)

2) F. Demichelis, G. Kaniadakis, A. Tagliaferro, E. Tresso and P. Rava
   Phys. Rev. B 37, 1231 (1988)

3) F. Demichelis, G. Kaniadakis, A. Tagliaferro, E. Tresso and G. Della Mea
   J. Appl. Phys. 64, 721 (1988)

4) F. Demichelis, G. Kaniadakis, A. Tagliaferro and E. Tresso
   Int. Journ. of Modern Phys. B 2, 237 (1988)

5) F. Demichelis, G. Kaniadakis, A. Tagliaferro and E. Tresso
   Solid State Comm. (in press)

6) J.C. Phillips
   Phys. Rev. Lett. 42, 1152, (1979)
   J. Non Cryst Solids 43, 37, (1981)

7) M.F. Thorpe
   J. Non Cryst. Solids 57, 355, (1983)

8) B. Dischler, R.E. Sah, P. Koidl, W. Fluhr and A. Wokann
   Proc. 7th Int. Symp. on Plasma Chemistry (ed. C.J. Timmermans - IUPAC Subcommittee of Plasma Chemistry, Eindhoven (1985)) p. 45

9) G. Kaniadakis, C.F. Pirri and E. Tresso
   Present book

Annealing effects on properties of a-CSiGe:H alloys.

G. Kaniadakis, C.F. Pirri, E. Tresso

Dipartimento di Fisica Politecnico di Torino (Italy)

Introduction.

The addition of carbon in the most studied binary alloys a-SiGe:H and a-SiSn:H has proved the remarkably influence on the properties of these alloys, due to the unusual structure of the carbon.

The ternary alloys produced by addition of carbon in a-SiGe:H and a-SiSn:H have been investigated (1,5).

The first results on the deposition condition and characterization of a-C:SiGe:H alloys are reported in references (4,5).

In this paper results on optical and structural properties of a-SiGe:H samples as deposited and after annealing are reported. Such results are complementary to the results appearing in another paper in this Proceeding.

Deposition conditions and characterization.

The films were deposited by Sputter Assisted Plasma Chemical Vapour Deposition (SAPCVD). The magnetron cathode is Ge (purity 99.999%) and the process take place in a mixture of

SiH_4, CH_4 and Ar. The Ar and SiH_4 flows were kept constant while the CH_4 flow was varied in the range 0-10 sccm. The deposition conditions are reported in Reference 5.

Transmittance and reflectance measurements were performed in the wavelength range 0.2-1.5 μm, by means of a Perkin Elmer UV-visible-NIR Lambda 9 spectrophotometer.

Reflectance measurements from both the air and the substrate sides of the film were compared with the reflectance of a standard mirror calibrated at the NAtional Bureau of Standards (NBS).

The optical properties, that is refractive index, extinction coefficient, absorption coefficient and optical gap were extracted from transmittance and reflectance measurements following the procedure described in Reference 6.

I.R. spectroscopy measurements were made in the absorption mode using a single beam Perkin Elmer 1710 Fourier spectrophotometer between 400 and 4000 cm^{-1}.

All the measurements were performed on the same samples as deposited and after annealing at 250° C.

Results.

Fig.1 shows the transmittance and reflectance of an a-SiGe:H and two a-CSiGe:H films. The films are labeled SiGe followed by a number indicating the at % of carbon (5).

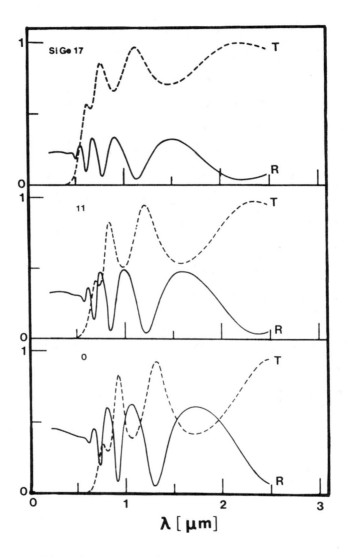

Fig. 1 - Transmittance and reflectance behaviour for a-CSiGe:H films with different carbon content.

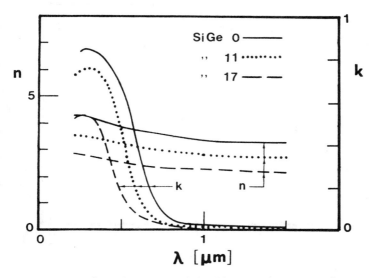

Fig. 2 - n and k dispersion trend for the same film of Fig. 1.

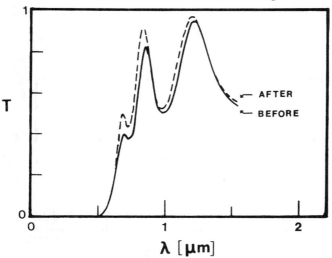

Fig. 3 - Transmittance spectra of SiGe 11 before and after annealing at 250°C.

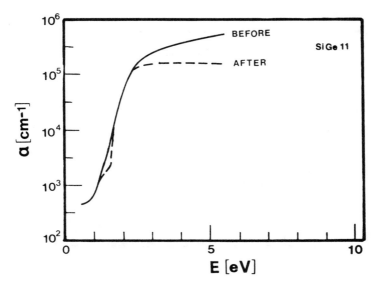

Fig. 4 - Absorption coefficient before and after annealing for the sample of Fig. 3.

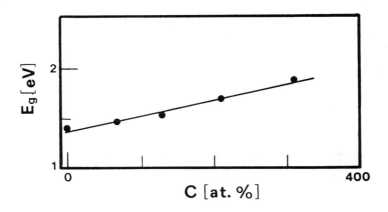

Fig. 5 - Energy gap values vs. carbon atomic percentage.

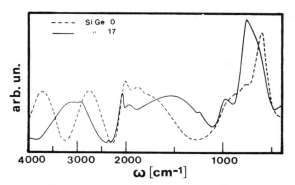

Fig. 6 - I.R. absorption spectra for
a-SiGe:H and a-CSiGe:H films.

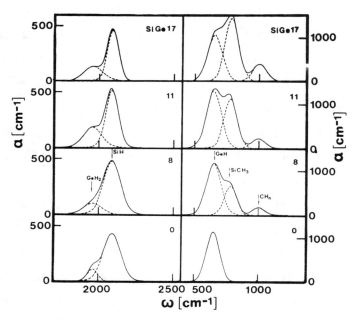

Fig. 7 - Deconvoluted I.R. peaks for a-CSiGe:H
with different carbon content.

It can be readily observed that the presence of carbon improves remarkably the transmittance of the films. Results on refractive index n and extinction coefficient k are reported as a function of the wavelength in fig. 2. As carbon content increases the refractive index curves are shifted to lower values, and the extinction coefficient show lower peak values.

Fig. 3 report the transmittance of the sample SiGe11 before and after annealing showing an improving in the transmittance in annealed sample.

We have no seen remarkable difference in the index of refraction and the extinction coefficient, while the absorption coefficient α has changed as is shown in fig. 4. We observe a remarkable reduction in the low energies corrisponding to the state into the gap and no difference in the fondamental absorption region.

The behavior of Eg as a function of the square of atomic % of carbon is shown in fig. 5. The gap of the samples is not changed after annealing.

Fig. 6 contains two I.R. spectra for films with no carbon and 17 at% carbon.

From I.R. spectra the absorbtion coefficient has been deduced taking into account the influence of interference fringes and absorption due to silicon substrate. Figs 7 shows the deconvoluted I.R. peaks in the range 400 to 1200 cm^{-1} for samples deposited under different CH_4 flows. We observe the presence in all the samples of the Si-H stretch vibration (2100 cm^{-1}). The band at 1800-2000 cm^{-1} assigned to GeH_2 shows a double structure (7) corresponding to the respective absorption of Ge-H groups (1875 cm^{-1}) and GeH_2 groups (1960 cm^{-1}) stretching modes.

The peak at 650 cm^{-1} seems to be attributed to Si-H and Ge-H wagging modes (8). The peak at 780 cm^{-1} is attributed to $Si-CH_3$ rocking or wagging modes and that at 1000 cm^{-1} to CH_2 rocking and/or wagging modes (9).

Fig. 8 shows the peaks of Si-H and GeH_2 vibrations before and after annealing for the sample SiGe11; we can observe that the Si-H peak is unchanged while a slow decrease is present in GeH_2 peak. A more remarkable change is shown in the behavior of the Ge-H and CH_2 peaks for the same sample as we can see in fig. 9.

The same trend is observed in all the samples more notably as the carbon content increases.

From these results we deduce a decrease of Ge-H and an increase in CH_2 bonds, while the Si-H bond remains unchanged.

Fig. 10 shows the variation in % of the intensities of the different bonds defined as

$$S = \int \frac{\alpha(\omega)}{\omega} d\omega$$

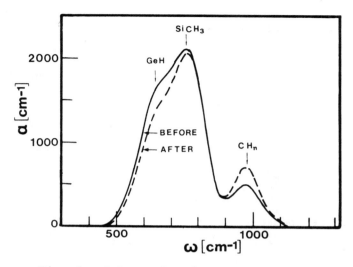

Fig. 8 – I.R. peaks of sample SiGe 11 before and after annealing at 250°C.

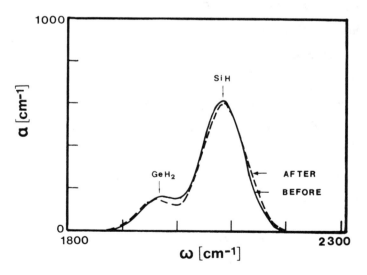

Fig. 9 – I.R. peaks for sample SiGe 11 before and after annealing at 250°C.

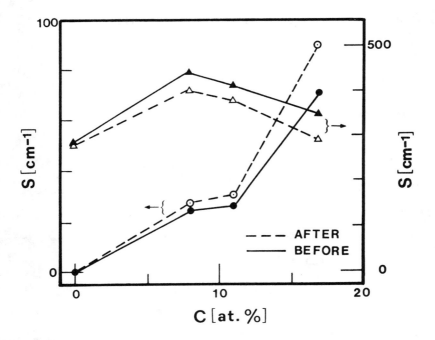

Fig. 10 - I.R. integrated intensities of GeH (△,▲) and CH (○,●) peaks before and after annealing as a function carbon content.

where $\alpha(\omega)$ is the absorption coefficient, as a function of carbon content, of the two Ge-H and CH_2 peaks.

After anmnealing the hydrogen lost from germaniumis captured from carbon. These experimental results confirm the higher binding energy of hydrogen to silicon with respect to germanium and to carbon with respect to silicon.

Conclusions.

From the present results few general trends emerge.
Addition of carbon in a-SiGe:H alloy improves the optical properties and give the possibility of a wide range of optical gap. The annealing of the samples leaves the optical properties pratically unchanged.
The I.R. spectra show different intensities of the Ge-H and CH_2 vibrations before and after annealing indicating a transfer of hydrogen from Ge-H to CH_2 bonds. The Si-H intensities remain unchanged. The increase of CH_2 bonds in annealed samples can explain the unusual behavior of the temperature dependence of the dark conductivitry as it is shown in another paper in this proceeding.

References.

1) F. Demichelis, G. Kaniadakis, A. Tagliaferro, E. Tresso and P. Rava - Phys. Rev. B 37,1231 (1988).

2) F. Demichelis, G. Kaniadakis, A. Tagliaferro, E. Tresso P. Rava and G. Della Mea - J. Appl. Phys. 64,721 (1988).

3) F. Demichelis, G. Kaniadakis, A. Tagliaferro and E. Tresso - Int. J. Mod. Phys. B 2,237 (1988).

4) F. Demichelis, G. Kaniadakis, A. Tagliaferro, E. Tresso and P. Rava - Proc. 3rd Int. PVSEC, Tokyo, 313 (1987)

5) F. Demichelis, G. Kaniadakis, A. Tagliaferro, E. Tresso, G. Della Mea and A. Paccagnella - Solid State Comm. (in press)

6) F. Demichelis, G. Kaniadakis, A. Tagliaferro and E. Tresso - Appl. Opt. 26,1737 (1987)

7) Y. Katayama and T. Shmiada - Jpn. J. Appl. Phys. 19, Suppl. 14-2,115 (1980)

8) Y. Catherine and G. Turban - Thin Solid Films 70,101 (1980)

9) H. Wieder, M. Cardona and C.R. Guarnieri - Phys. Stat. Sol. (b) 92,99 (1979)

PART II
DEVICES

Transport in Schottky barrier structures on amorphous semiconductors

P. J. McElheny and S. J. Fonash

Center for Electronic Materials and Processing and the Engineering Science Program,
The Pennsylvania State University, University Park, Pennsylvania 16802

Abstract

Using a first-principles computer model, we demonstrate that there is a rich variety of transport behavior possible for Schottky barrier devices made on amorphous semiconductors such as hydrogenated amorphous silicon. It is found that the reverse current can vary between being generation limited for large barriers and barrier limited for small barriers. The forward current is found to be dominated by three different types of transport. Two of these are the expected recombination current at low forward bias and thermionic emission or thermionic emission-diffusion at moderate forward bias. The third transport mechanism is space charge limited current which is found to dominate at far forward bias. Contrary to the usual concept of space charge limited current, however, transport in this latter voltage regime is found to be dependent on both bulk and contact properties. Further, it is demonstrated that a contact-dependent conductivity modulation occurs in this space charge limited current regime. This renders extracting the built-in potential, bulk Fermi level position, and gap state information problematic in this far forward bias region.

Transport in Schottky barrier structures on amorphous semiconductors

P. J. McElheny and S. J. Fonash

Center for Electronic Materials and Processing and the Engineering Science Program,

The Pennsylvania State University, University Park, Pennsylvania 16802

Introduction

In this paper we use a first principles computer model, with no *a priori* assumptions about what is controlling transport, to generate the J-V-T characteristics of amorphous semiconductor Schottky barrier devices. For definitiveness it is assumed that the amorphous semiconductor is a-Si:H. We generate the J-V behavior as a function of temperature just as one would do experimentally. Using this model we determine precisely what is controlling transport in various bias regimes. We therefore establish when a simple crystalline semiconductor-type of approach to analyzing a-Si:H Schottky barrier J-V data is valid and when it is not. Obviously, in the latter case, interface and bulk material parameters extracted using the simple transport equations of crystalline semiconductor theory are fallacious. We then end with an assessment of experimental transport analyses made on Schottky barrier devices.

Computer Model

We have described our general computer model in detail elsewhere.[1,2] Briefly, the model simultaneously solves Poisson's equation, the Shockley-Read-Hall recombination-generation equation, and the continuity equations for electrons and holes. We convert these equations to a set of difference equations and solve them using the Newton-Raphson technique. Our numerical model is similar to that of Schwartz *et al.*[3] and Hack and Shur[4] except that we employ more general boundary conditions with surface recombination speeds and surface barrier heights as we have discussed previously.[1,2,5] In addition, unlike our previous reports and unlike other published models[3,4] we no longer use the Taylor-Simmons approximation (T=0K) to the occupation probability of the localized gap states.[6] Instead, in this work we have numerically integrated across the bandgap using occupation probabilities that account for the temperature to determine the trapped carrier populations in the gap states and the recombination rate through these states. Hence, our computer model is a first principles approach that properly accounts for the temperature and properly accounts for transport across interfaces. Our model has embodied in it, from first principles, the complete spectrum of transport mechanisms from space charge limited current to thermionic emission-diffusion Schottky barrier controlled current.

Discussion

We begin this discussion of what can be learned from a first-principles computer analysis of transport in a-Si:H Schottky barrier structures by showing in Figures 1-3 computer-generated J-V characteristics for various Schottky barriers on a-Si:H. These devices have an n+ back contact and, in general, are described by the insert in Figure 1. The density of gap states energy dependence employed in this modelling is that which we have published elsewhere.[1,2] To be specific, we use exponential densities coming out of each band merging with a flat mid-gap region density of states G_{MG}. The states from the valence band edge to an energy E_{DA} in the mid-gap range are donor-like; the states from E_{DA} to the conduction band are acceptor-like. All these devices are seen to have Schottky barrier heights of ϕ_{bo}=1.2 eV or ϕ_{bo}=0.9 eV. The devices of Fig. 1 and Fig. 2 have a mid-gap density of states of G_{MG}=5X10^{15} eV^{-1}cm^{-3} and a length of 1.1 µm and 3.1 µm, respectively. The devices of Fig. 3 have a mid-gap density of states of G_{MG}=5X10^{16} eV^{-1}cm^{-3} and a length of 3.1 µm. We list all the input parameters for these devices in Table I. The definitions of the parameters are the usual standard definitions; we have used them in our previous reports.[1,2] Three important features that we will discuss may immediately be noted from these curves of Figs. 1-3. The first is the manner in which the ideality factor n, defined as $J=J_0\exp(-qVF/nkT)$, varies as a function of barrier height and voltage. The second is the saturation behavior, or lack thereof, of the reverse current. Finally, and most importantly, the third is the general lack of convergence of the J-V characteristics at far forward bias.

The first point we will address concerns the ideality factor. Treating a computer generated J-V characteristic as if it were experimental data, we can compute an "experimental" ideality factor n for different regions of the J-V curve. By treating computer generated J-V-T data as if it were experimentally determined, we can also extract the "experimentally-determined" activation energy for the different regions of the J-V curve. This experimentally determined activation energy W is defined by the usual manner by $J=J_0\exp(-W/kT)$. In addition, we know from the computer analysis what transport mechanisms are really controlling the current at a given voltage. When we use this approach, analysis of our computer results shows that, at low bias, the current is dominated by recombination. This occurs in voltage regions where the "experimentally determined" n-factor is approximately 2. For a barrier height of $\phi_{bo}=1.2$ eV, this low bias, recombination-dominated regime is seen from Figs. 1-3 to exist for voltages ≤0.2 V. Fig. 4 presents the results of "experimentally determining" the activation energy W for the devices of Fig. 1 using the temperatures of T=300K, T=325K, T=350K, and T=375K. One would expect $W=E_G/2 - V/2$[7] for this n≈2 region; hence, from Fig. 4 we "experimentally" deduce that E_G~1.7 for the devices of Fig. 1 whereas we note from Table I that $E_G=1.72$ eV was actually used in the computer model. As we see, the agreement between the actual E_G value and that deduced "experimentally" from this n=2 region is good.

As the forward bias is increased, analysis of the computer results for the $\phi_{bo}=1.2$ eV structures shows that the current is becoming influenced by the Schottky barrier. In this region of voltages for these devices, the "experimentally determined" n-factor moves towards unity, as one would expect. From Figs. 1-3 it is seen that this region exists for voltages in the range from 0.2-0.8 V for $\phi_{bo}=1.2$ eV. Above voltages ≥0.8 V the computer analysis shows that the ideality factor concept no longer applies for these $\phi_{bo}=1.2$ eV devices since they enter a space charge limited current (SCLC) regime, as we will discuss shortly. It is interesting to note how this shift from one transport controlled regime to another varies with different device parameters. From Fig. 1, we see that as we reduce the barrier to 0.9 eV for the 1.1 μm device, the trend shifts to the point where the device enters the space charge limited current (SCLC) regime before it reaches the Schottky barrier regime. Thus, the experimentalist must be cautious when trying to extract a barrier height from experimental J-V data. As this example shows for the ϕ_{bo}=0.9 eV case there is no voltage range where thermionic emission theory, or any of its modified forms such as thermionic emission-diffusion theory, holds.

Fig. 4 explores the experimental approach of extracting ϕ_{bo} from voltage regions where n approaches unity. The "experimentally determined" activation energy W for the voltage region where one would expect W=ϕ_{bo}-V may be seen in Fig. 4 for the devices of Fig. 1. Results for ϕ_{bo}=1.0 and 1.1 eV are included with those of ϕ_{bo}=1.2 and 0.9 eV. As may be seen, the "experimentally determined" ϕ_{bo} value of 1.2 eV is similar to the actual model input over a range of about .3 volts. Similar ranges exist for ϕ_{bo}=1.1 and 1.0 eV, while this range does not exist for ϕ_{bo}=0.9 eV. Therefore, as long as the ideality factor is close to 1, one can extract a reasonable value for ϕ_{bo}.

As we have noted, a second point of interest in these curves of Figs. 1-3 is the saturation behavior of the reverse bias current. Analyzing our computer results to determine what transport mechanism is actually dominating, we find that generation dominates for the larger barrier heights and thermionic emission dominates for the smaller barrier heights. The currents do not saturate in reverse bias because at low reverse biases, only a small portion of the bulk of the device is involved in the generation process due to the large density of states in the gap. As the reverse bias is increased, more of the bulk of the device is involved in the generation process, thus leading to an increase in current. Therefore, the reverse current will eventually saturate even when it is controlled by generation. This will occur at lower reverse biases both for thinner devices and for better quality material. It is interesting to note from Fig. 4 that the activation energy deduced from this "experimentally determined" activation analysis is a constant in reverse bias. However, it is seen that this constant need not be $E_G/2$ or ϕ_{bo} due to the mix of over-the-barrier transport and generation at intermediate barrier heights. Recalling how the reverse bias current is dominated by the generation process for large barrier heights and by the barrier height for smaller barrier heights, it is evident that at intermediate barrier heights it is a mixture of these two processes.

The third and most important feature of these figures is how the far forward bias current does not, in general, converge to one bulk-controlled J-V curve for different barrier heights. This immediately tells us that the usually assumed generalization that transport in this voltage regime is controlled by bulk material properties is invalid. Analysis of the computer results for this voltage regime shows that the current in this case is controlled by drift across the bulk i-a-Si:H layer. It further shows that the free electron population, which is carrying this current, is a function of the electrostatic field in the bulk. Further, this field is established by the subset of electrons which is trapped in gap states. Hence, by definition, this far forward voltage regime is, for the devices of Figs. 1-3, controlled by space charge limited current (SCLC). The devices of Figs. 1-3 begin to enter this far forward bias regime when the applied bias equals the built-in potential, which explains why these curves intersect at different voltages in their J-V's depending on their barrier heights. The short devices (length = 1.1 μm) of Fig. 1 do tend to show convergence to one SCLC characteristic for this far forward bias region whereas the longer devices of Figs. 2 and 3 definitely do not. That is, for the devices of Fig. 2 and Fig. 3 the SCLC characteristics vary with Schottky barrier height. Since shorter devices require more trapped electrons to establish the field, one can understand how the SCLC behavior converges to a characteristic of the onset of the conduction band tail for shorter devices. This SCLC behavior has been seen to reduce to ohmic behavior for relatively long devices of poor quality material. For example, we observe Ohmic behavior for $G_{MG}=5 \times 10^{17}$ cm^{-3}eV^{-1} and a device length of 5 μm. Therefore, unless one has a long device or one of very poor quality material, the far forward bias will be controlled by space charge limited current.

The fact that the SCLC regime is, in general, dependent on contact properties and thickness as well as bulk properties has important implications for the experimental determination of material parameters. It is this far forward bias regime that is often used to experimentally extract information on the built-in potential, bulk Fermi level position, and gap state energy distribution. Unfortunately, the computer results show that the barrier height and thickness of a device interplay, in general, to create a conductivity modulation effect in this bias regime and, hence, it is impossible to extract these parameters in a simple manner. That is, a virtual cathode must be formed when a device enters the SCLC regime, and the amount of trapped electrons necessary to create the virtual cathode is dependent on the barrier height and i-layer thickness. Our computer analysis shows the larger the barrier height and the thinner the device, the more i-layer trapped charge is required. Since this also results in more of the far-fewer free electrons, there is a conductivity modulation effect which, as Figs. 1-3 show, is enhanced by a larger barrier height and thinner devices. We believe this may explain why others have reported conductivity increases in thinner devices.[8] We will discuss this phenomenon of conductivity modulation which leads to barrier-dependent SCLC characteristics more fully in a later publication.

Conclusions

In conclusion, our first principles computer analysis of transport mechanisms in Schottky barrier devices on amorphous semiconductors has produced results that help both to confirm and question certain interpretations of experimental results. In particular, our analysis demonstrates that a reasonable value for barrier height can be extracted when the ideality factor $n \sim 1$, and when $n \sim 2$, one can extract a good measure of half the recombination bandgap. However, the far forward bias regime of these Schottky barrier devices has been shown to be space charge limited current that depends not only on bulk parameters but also on contact parameters. In addition, the reverse bias is shown to be primarily a mix of over-the-barrier current and generation, thus making it difficult to extract parameters. Therefore, we suggest that one extract information on the barrier height and bandgap from the low to moderate forward bias regime only if the barrier height is approximately 1.0 eV or larger; otherwise, a reasonable number may not be obtainable due to the early onset of SCLC. In addition, since the far forward bias regime is a complicated function of the bulk, contacts, and device length, more analysis is necessary before useful information can be extracted.

We would like to acknowledge the support of the Electric Power Research Institute in funding this work under contract number 8001-03.

References

1. P. J. McElheny, J. K. Arch and S. J. Fonash, Appl. Phys. Lett. **51**, 1611 (1987).
2. P. J. McElheny, J. K. Arch and S. J. Fonash, J. Appl. Phys. **64**, 1254 (1988).
3. R. J. Schwartz, J. L. Gray, G. B. Turner, D. Kanani, and H. Ullal, in *Proceedings of the 17th IEEE Photovoltaic Specialists Conference* (IEEE, New York, 1984), pp. 369-373.
4. M. Hack and M. Shur, J. Appl. Phys. **58**, 997 (1985).
5. S. J. Fonash, Solar Cell Device Physics (Academic, New York, 1981).

6. G. W. Taylor and J. G. Simmons, J. Non-Cryst. Solids **8-10**, 940 (1972).

7. S. M. Sze, Physics of Semiconductor Devices (John Wiley & Sons, New York, 1981).

8. E. V. Grekov and O. G. Sukhoruukov, Sov. Phys. Semicond. **22**, 457 (1988).

Figure 1. Numerical solutions of the dark current density vs. voltage for various barrier heights. The input parameters for the model are listed in Table I.

Figure 2. Numerical solutions of the dark current density vs. voltage for various barrier heights. The input parameters for the model are listed in Table I with one difference: the mid-gap density of states is increased to 5X1016.

Figure 3. Numerical solutions of the dark current density vs. voltage for various barrier heights. The input parameters for the model are listed in Table I with one difference: the length of the device is increased to 3.1 μm.

Figure 4. Numerical solutions of an activation energy analysis performed for the device of Table I with various barrier heights.

TABLE I. Input parameters for Figures 1 and 4.

$L=1.1\ \mu m$	$G_{D0}=G_{A0}=10^{21}\ cm^{-3}/eV^{-1}$	$E_A=30\ meV$
$E_G=1.72\ eV$	$G_{MG}=5\times10^{15}\ cm^{-3}/eV^{-1}$	$E_D=50\ meV$
$N_{NET}=0\ cm^{-3}$	$S_{n0}=S_{p0}=10^7\ cm/s$	$E_{DA}=0.86\ eV$
$\mu_n=10\ cm^2/V\ s$	$\sigma_c=10^{-15}\ cm^2$	
$\mu_p=1\ cm^2/V\ s$	$\sigma_n=10^{-16}\ cm^2$	

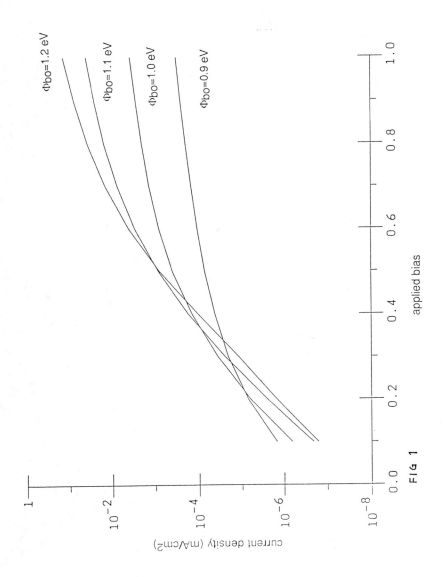

FIG 1

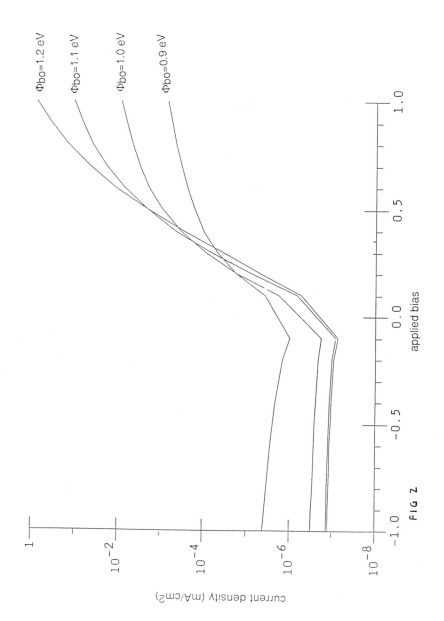

FIG 2

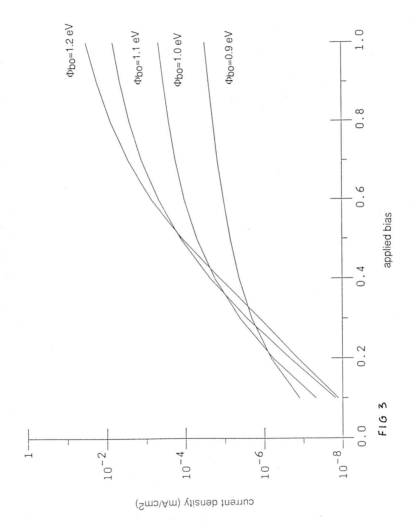

FIG 3
ln(J) vs. V for different barrier heights

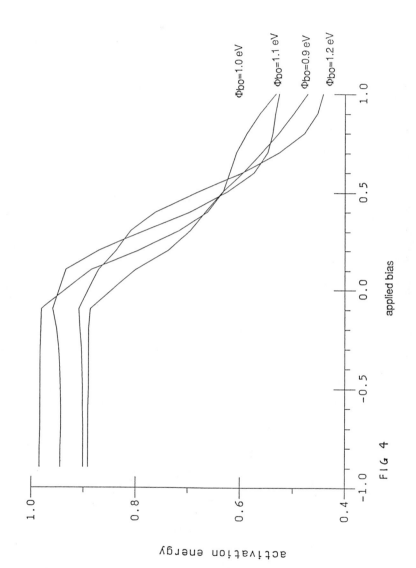

FIG 4

GENERATIONS AND RECOMBINATIONS IN a-Si IN TIME-VARIABLE CONDITIONS

Joze FURLAN, University of Ljubljana, Yugoslavia

ABSTRACT:
In single crystal Si the density of localized states in the bandgap is relatively low so that the trapped charge carrier densities and their time variations can be neglected. On the contrary, this is not in general permitted in treating the a-Si having high and continuous distribution of gap states. In time-variable conditions any change of free charge carriers is generally accompanied by a corresponding charge of trapped charge carriers. The resulting time variation of occupancy distribution within the mobility gap reflects back the rate of change of free charge carriers creating unequal rates dn/dt and dp/dt. Based on SRH generation-recombination model the rates of change of free and trapped charge carrier densities are formulated by a set of four time dependent differential equations. An illustrative insight into time variations of free and trapped electrons and holes in dynamic nonequilibrium conditions is presented showing the transient response of light generated carriers in undoped and doped a-Si approaching their thermal equilibrium densities after removing the illumination.

1. GENERATION-RECOMBINATION APPROACH IN a-Si

The principal material property affecting generation recombination rate in a-Si is its large and continuous distribution of gap states in the mobility gap. As a consequence, some of assumptions used in dealing with crystalline silicon can not be applied to a-Si. These are mainly the following:

- since the trapped charge carrier concentrations are large compared to free carrier densities, $n_t \gg n$ and $p_t \gg p$, the time variations dn_t/dt and dp_t/dt may not be neglected (except in the special case of steady-state conditions),

- approximations using low level excitation can not be applied because the nonequilibrium densities of free charge carriers are generally much higher than their thermal equilibrium values, $n \gg n_o$ and $p \gg p_o$.

The net rates of change of charge carriers can be described by a pair of continuity equations

$$\frac{dn}{dt} = G_n - R_n + G_L - \Delta\phi_n, \qquad \frac{dp}{dt} = G_p - R_p + G_L - \Delta\phi_p \qquad (1)$$

accompanied by two additional expressions governing the concentration of charge in localized states

$$\frac{dn_t}{dt} = \frac{dn_{tA}}{dt} = R_{nA} + G_{pA} - R_{pA} - G_{nA}$$
$$\frac{dp_t}{dt} = -\frac{dn_{tD}}{dt} = R_{pD} + G_{nD} - R_{nD} - G_{pD} \qquad (2)$$

The assumptions adopted in writing these equations are the following:

- light generations G_L transfer electrons only from the valence band into the conduction band,

- charge carrier flow is confined only to conduction and valence band and is neglected inside the mobility gap.

Thermal generation-recombination rates in a-Si in time varying excitations are obtained following SRH generation-recombination model /1/. The four electron and hole capture and emission processes in the differential energy range dE at energy E in the mobility gap, shown in Fig.1, contribute to the rates of change of free and trapped charge carrier densities as follows

$$d\frac{dn}{dt} = \sigma v_{th}(n_i^2 - pn)\left[\frac{g_D dE}{R(n+n_1)+p+p_1} + \frac{g_A dE}{n+n_1+R(p+p_1)}\right] + \frac{d\frac{dp_t}{dt}}{1+\frac{p+p_1}{R(n+n_1)}} - \frac{d\frac{dn_t}{dt}}{1+\frac{R(p+p_1)}{n+n_1}} \qquad (3)$$

$$d\frac{dp}{dt} = \sigma v_{th}(n_i^2 - pn)\left[\frac{g_D dE}{R(n+n_1)+p+p_1} + \frac{g_A dE}{n+n_1+R(p+p_1)}\right] - \frac{d\frac{dp_t}{dt}}{1+\frac{R(n+n_1)}{p+p_1}} - \frac{d\frac{dn_t}{dt}}{1+\frac{n+n_1}{R(p+p_1)}} \qquad (4)$$

$$d\frac{dn_t}{dt} = \frac{\sigma v_{th}}{R}\left[(n+Rp_1)g_A dE - (n+n_1+R(p+p_1))dn_t\right] \qquad (5)$$

$$d\frac{dp_t}{dt} = \frac{\sigma v_{th}}{R}\left[(p+Rn_1)g_D dE - (R(n+n_1)+p+p_1)dp_t\right] \qquad (6)$$

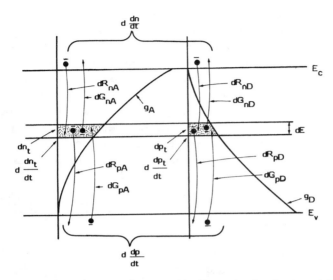

Fig. 1. Thermal generation-recombination rate due to gap states in the differential energy interval in the mobility gap

where g_D and g_A represent continuous donorlike and acceptorlike gap states distributions and the relations between capture cross-sections are given by

$$\sigma = \sigma_{nD} = \sigma_{pA} \qquad \frac{\sigma_{nD}}{\sigma_{nA}} = \frac{\sigma_{pA}}{\sigma_{pD}} = R \gg 1$$

The first term on the right-hand side in equations (3) and (4) is the well known generation-recombination rate resulting from steady-state analysis assuming that the trapped carrier densities do not change with time /2/,/3/. In this particular case the generation-recombination rate of free electrons equals that of free holes.

In time-variable conditions however, any change of free charge carriers is generally accompanied by certain changes of trapped carrier densities. The resulting time variations of occupancy destributions within the mobility gap reflect back the rates of change of free charge carriers creating thus in general unequal rates dn/dt and dp/dt.

After describing the contributions to generation-recombination rates from gap states located in an energy range dE, the total rates of change of free and trapped charge carriers are obtained by integrating over all contributions of gap states throughout the mobility gap between E_v and E_c. To enable analytical integration the same approximation adopted in steady-state analysis may be applied. It is based on the fact that in excess carrier nonequilibrium conditions ($np > n_i^2$) only the gap states located between electron and hole trap quasi-Fermi levels contribute to the captured charge carrier densities, n_t and p_t, as well as to the net generation-recombination rate /2/. Using this approximation, the net rates of change of charge carriers are given by

$$\frac{dn}{dt} = -\sigma v_{th} pn \left[\frac{1}{Rn+p} \int_{E_{1D}}^{E_{2D}} g_D dE + \frac{1}{n+Rp} \int_{E_{1A}}^{E_{2A}} g_A dE \right] + \frac{R_n}{Rn+p} \frac{dp_t}{dt} - \frac{n}{n+Rp} \frac{dn_t}{dt} \quad (7)$$

$$\frac{dp}{dt} = -\sigma v_{th} pn \left[\frac{1}{Rn+p} \int_{E_{1D}}^{E_{2D}} g_D dE + \frac{1}{n+Rp} \int_{E_{1A}}^{E_{2A}} g_A dE \right] + \frac{R_p}{n+Rp} \frac{dn_t}{dt} - \frac{p}{Rn+p} \frac{dp_t}{dt} \quad (8)$$

$$\frac{dn_t}{dt} = \sigma v_{th} \frac{n+Rp}{R} \left[\frac{n}{n+Rp} \int_{E_{1A}}^{E_{2A}} g_A dE - n_t \right] \quad (9)$$

$$\frac{dp_t}{dt} = \sigma v_{th} \frac{Rn+p}{R} \left[\frac{p}{Rn+p} \int_{E_{1D}}^{E_{2D}} g_D dE - p_t \right] \quad (10)$$

where the integration limits equal electron and hole trap quasi-Fermi levels defined as shown in Fig.2 /3/.

$$\begin{aligned}
E_{1A} &= E_i - kT \ln(p+n/R)/n_i = E_v - kT \ln(p+n/R)/N_v, \\
E_{2A} &= E_i + kT \ln(n+Rp)/n_i = E_c + kT \ln(n+Rp)/N_c, \\
E_{1D} &= E_i - kT \ln(Rn+p)/n_i = E_v - kT \ln(Rn+p)/N_v, \\
E_{2D} &= E_i + kT \ln(n+p/R)/ni = E_c + kT \ln(n+p/R)/N_c.
\end{aligned} \quad (11)$$

It can be easily shown that the net rate of negative charge carriers equals the rate of all positive charge carriers

$$\frac{dn}{dt} + \frac{dn_t}{dt} = \frac{dp}{dt} + \frac{dp_t}{dt} = -\sigma v_{th}(np_t + pn_t) \quad (12)$$

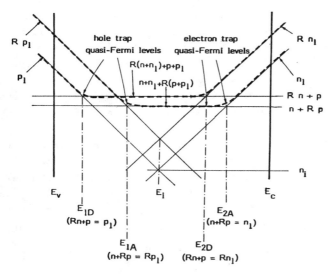

Fig. 2 Graphical representation of terms defining trap quasi-Fermi levels

ensuring that the conservation of charge provides the electrical neutrality during time varying conditions. Expressions (7) to (10) together with integration limits given by eqs. (11) is a set of approximate equations describing the rates of change of charge carriers in a-Si. Solutions of these equations giving the time dependence of carriers depend primarily on the assumed localized states distribution, on the density of purposedly added impurities and on the method of thermal nonequilibrium excitation.

2. EXAMPLES OF TRANSIENT RESPONSE IN a-Si

A) Transient response in an undoped a-Si

As a first example the transient response in an undoped a-Si following the removal of the illumination will be examined. To simplify the interpretation of events during transients a symmetrical double-exponential distribution of gap states given by

$$g_A = A \exp(E-E_C)/W, \qquad g_D = A \exp(E_V-E)/W$$

will be assumed with the consequence of equal electron and hole densities ($n=p$, $n_t=p_t$) approaching their thermal equilibrium concentrations as shown in Fig.3.a.

During the illumination ($t<0$) there are large populations of excess carriers. In the moment of breaking the illumination ($t=0$), the occupation of states is still unchanged. Consequently, the thermal rates of change of carrier densities at $t=0$ are the same as before light removal. The absence of light generations results in large decreasing rates of free charge carriers approaching their thermal equilibrium densities.

After a short elapsed time ($t>0$), the excess concentrations of free electrons and holes have slightly dropped, lowering also occupation probabilities. As a consequence the trapped excess carrier concentrations have also begun to decay towards their thermal equilibrium values. Any change of trapped carrier

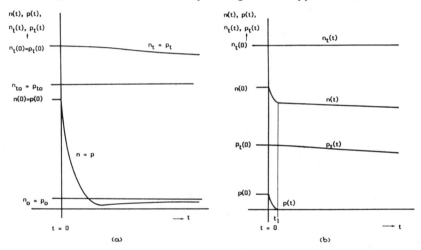

Fig. 3. Transient response of charge carriers
a) in undoped a-Si in case of symmetrical gap states distribution
b) in N-type doped a-Si.

concentration reflects back the generation-recombination mechanisms. Here it should be noted that even in case of symmetrical donorlike and acceptorlike gap states distributions, the resulting equal rates $dn_t/dt = dp_t/dt$ do not influence the rates of free electrons and holes in the same extent. Due to different capture cross-sections of carriers in donorlike and acceptorlike centers the transitions of conduction electrons are more sensitive to the trapped hole rate dp_t/dt and less to the trapped electron rate dn_t/dt. Consequently, the recombination rate of free electrons dn/dt in the presence of decreasing density of trapped holes is higher than it would be in case of constant trapped carrier density. Similar argument applies to the decreasing rate of free holes which is affected primarily by the decreasing rate of trapped electrons in acceptorlike states.

After certain time free carrier densities approach their thermal equilibrium values, yet in the presence of very high excess concentrations of trapped carriers. Even when free carrier concentrations reach their thermal equilibrium values the decreasing rates dp_t/dt and dn_t/dt cause a further decrease of free electron and hole densities. After reaching the minimum point in free carrier concentrations, both free and trapped charge carrier densities monotonously approach their thermal equilibrium values by transferring the excess electrons from acceptorlike centres over conduction and valence band into donorlike centres compensating free excess concentration of trapped holes.

The described transient response can be solved analytically using the set of derived eqs. (7) to (10). The resulting time constant of decaying free carriers is very short and that of trapped carriers is very long

$$\tau_n = \tau_p = \frac{1}{2\sigma v_{th} n_t(0)} \qquad \tau_t > \frac{1}{\sigma v_{th} n_i}.$$

B) Transient response in N-type doped a-Si:

Starting with the physical discussion of effects during transient represented by a typical time response shown in Fig.3,b, the steady-state conditions during continuous illumination are selected such that

$$n_t \approx N_D >> P_t, \qquad n_t >> n >> p.$$

At the moment the light generations are turned off, the thermal recombination rate of free charge carriers remains unchanged resulting in a large decreasing rate of free electrons and holes, whereas the trapped carrier densities remain $dn_t/dt = dp_t/dt = 0$.

Free electron and hole densities are rapidly decreasing with a very short time constant while on the contrary the densities of trapped electrons and holes do not change considerably.

Within a short time this situation is changed drastically. Since the majority carrier free electron density predominates greatly over minority hole density, equal rates dn/dt and dp/dt result in a very large relative drop of hole density compared to a small relative change of electron density. As a consequence, the density of free holes is becoming very small thus limiting also the rate dp/dt to a negligible amount. As the rate of the total negative charge must equal the rate of the total positive charge and because a direct transfer of trapped electrons into trapped holes is not likely, the excess electron density fills up empty donorlike states ($dn/dt \approx dp_t/dt$). This rate however represents a very slow process due to relatively low concentrations of both, free electrons n and empty gap states p_t. As a result both densities, n and p_t, are slowly decreasing towards their thermal equilibrium concentrations.

Using again the set of eqs.(7) to (10) with appropriate boundary conditions, the obtained initial time constant of free charge carriers

$$\tau_n = \tau_p = \frac{1}{\sigma v_{th} N_D}$$

is very low and the time constant of free electrons and trapped holes following the initial drop of free carriers

$$\tau_n = \tau_{pt} = \frac{1}{\sigma v_{th} n(t_1)}$$

becomes very large.

REFERENCES:
1. W. Shockley, W. T. Read, Phys. Rev., vol. 87, pp. 835-842, 1952.
2. M. Hack, M. Shur, J. Appl. Phys., vol. 54, pp. 5858-5863, 1983.
3. J. Furlan, S. Amon, MRS Symp. Proc., vol. 70, pp. 149-154, 1986.

COMPUTER MODELLING OF INTERNAL AND EXTERNAL
PROPERTIES OF a-Si:H SOLAR CELLS

Franc Smole, University of Ljubljana, Yugoslavia

ABSTRACT:

A computer program enabling the analysis of an illuminated p-i-n structure of a-Si:H solar cell is developed. Models in this program are arranged to fit closely experimentally established localized states distribution in the mobility gap of a-Si. The model of localized states distribution includes the tails of donorlike and acceptorlike states as well as dangling bond defect states having positive and negative correlation energies. Generations and recombinations are expressed following an expanded SRH approach. The developed computer program solves electron and hole continuity and transport equations as well as Poisson equation where free and trapped charge carrier densities are taken into account. Due to intricate expressions for trapped carrier densities, n_t and p_t, the usual procedures for linearization of Poisson equation is not possible and some specific solutions are applied. In order to improve the convergence of continuity equations the Bernoulli function is used in discretization procedure. The developed computer modelling was used for investigation of both, internal properties and external characteristics of a p-i-n a-Si solar cell under AM-1 illumination.

I. Introduction

Computer simulation of internal electrical behavior in semiconductor structures can be an important tool for the investigation and optimization of semiconductor devices. The accuracy of obtained results naturally depend on the way of describing physical material properties. In the treatment of a-Si:H layers and structures one of the most important parameters is certainly the distribution of gap states. In spite of extensive investigations all details affecting gap states distribution are not yet completely understood and therefore an exact model representing gap states and their effects can not be defined today. Some approximated models suitable for mathematical analysis were introduced. As an example, Hack and Shur have used extensively a double-exponential approximation of gap states /1/, shown in Fig. 1.

In our approach we have tried to use a model of gap

states which would better suit experimental observations. Naturally, this model must include localized donorlike states tailing in mobility gap from the valence band, acceptorlike states tailing in mobility gap from the conduction band and the localized states distributions pertaining to defect states. In the adopted model the tail states follow Amer and Jackson distribution /2/. The peaks of distributions due to dangling bonds having positive correlation energy were chosen symmetrically in the bandgap. The peaks of localized states due to T_3^+ and T_3^- centers were adjusted in accordance with theoretical interpretations of Adler /2,3/ and with measurements of Morigaki /4/. In addition, in the location of T_3^+ states, the DLTS measurements were taken into account, showing the minimum of states nearly 0.4eV below the conduction band. As a consequence the T_3^+ states were placed about 0.2eV below the conduction band edge. The T_3^- states are more distant from the valence band. According to some expectations these states extend even to the lower region of positively correlated states due to dangling bonds /5/. It can be further concluded that in the neighborhood of dangling bonds and other defects, as well as around Si-H bonds there is a greater deformation of Si-Si bonds introducing another tail of donorlike states having a lower concentration but extending deeper into mobility gap, as shown in Fig. 2. The presented model was adopted as a basis for computer analysis of the illuminated p-i-n a-Si solar cell.

2. Analysis of dark and illuminated a-Si:H layers

The correctness of the assumed model was verified by comparing calculated properties of undoped and homogeneously doped a-Si:H layers with experimental data. The Fermi level as a function of doping level was determined from the condition of electrical neutrality, i.e. $\rho = 0$, valid in homogeneously doped semiconductor in thermal equilibrium. In contrast with single crystal Si where space charge includes free electrons and holes and ionized donor and acceptor impurities, the amorphous Si is characterized by large additional concentrations of trapped electrons and holes residing on allowed energy states in the mobility gap. The contributions of these trapped carriers are large compared to free carrier densities and therefore may not be disregarded. The total space charge in a-Si is therefore given by

$$\rho = q (p - n + p_t - n_t + N_D^+ - N_A^-),$$

where n and p are free electron and hole concentrations, n_t and p_t are trapped charge carriers concentrations, and where

N_D^+ and N_A^- represent concentrations of active ionized donor and acceptor impurities. For Fermi level determination, as is evident from equation for space charge, concentrations n_t and p_t must be known. Concentration of trapped electrons n_t was calculated by integrating the product of acceptorlike states density and occupation probability f_{tA} over mobility gap. Concentration of trapped holes p_t was determined similarly by integrating the product of donorlike states and the probability $1-f_{tD}$ over mobility gap. Due to existence of several types of states $g_A(E)$ and $g_D(E)$ contributions for each of these states densities was calculated first. Here some difficulties were encountered when calculating contributions to n_t, p_t from deffect states such as dangling bonds which introduce two states in mobility gap. In this case multielectron theory was applied. Finally, total concentrations of trapped electrons and holes n_t and p_t were determined by summation of all contributions from different types of states.

Results of these calculations including the dependence of impurity concentrations and of free and trapped charge carrier concentrations on the Fermi level position are shown in Fig. 3. In all doped samples the square root relation between gap states density and impurity concentration was taken into account /6/,/7/.

Concentrations of free and trapped charge carriers in an illuminated a-Si:H depend strongly on generation-recombination mehanisms. In an electrically neutral homogenously doped a-Si:H out of thermal equilibrium, where the basic relations are $\rho(x)=0$ and $G+G_L=R$, the situation can be described by occupation probabilities using SRH generation-recombination approach. Computed dependence of free and trapped charge carrier densities on light generations G_L are shown in Fig. 4.

3. Computer modelling of illuminated a-Si:H p-i-n solar cell

Electrical properties in semiconductor devices are described by five equations. These are a pair of transport equations, two continuity equations and the Poisson equation. Solving the resulting system of nonlinear differential equations the characteristics of devices seen at their output terminals as well as internal properties inside the investigated devices can be derived.

In the following solar cell analysis the steady-state one dimensional case with infinite recombination velocities on both surfaces was assumed.

Due to difficulties encountered when calculating the derivatives of trapped carriers on potential the linearization of Poisson equation was determined using Taylor series expansion.

The discretization of continuity equations involved Bernoulli functions in transport equations for electrons and holes. The important advantage of this method is in exponential representation of charge carrier densities leading to better computation convergence and accuracy. Another advantage is elimination of integration of recombination term R_i inside of each segment as it is usually necessary in other approaches.

The impurity profile of the analyzed p-i-n a-Si:H solar cell structure is shown in Fig. 5. Light generations under AM-1 illumination were represented by Johnson's formulation dividing the solar spectral distribution into numerous spectral bands. Fig. 6 shows distribution of free electrons and holes for different values of external applied voltages. These concentrations are much lower than corresponding concentrations in crystalline silicon due to wider bandgap and to larger concentrations of trapped carriers in a-Si:H, as shown in Fig. 7. The distribution of space charge in p-i-n structure is shown in Fig. 8. The main contribution to this charge is coming from large localized states concentrations in the undoped layer. The negative space charge in p^+ layer and the positive charge in n^+ layer produce a strong electric field, shown in Fig. 9. It is known that this electric field is indispensable for effective operation of a-Si solar cells. The recombination rate in a-Si is very high particularily in the doped region, as it is shown in Fig. 10. The final outcome of the analysis is the I-U characteristics of the illuminated p-i-n solar cell, shown in Fig. 11.

References:

/1/ M.Hack and M.Shur, J.Appl.Phys. 58, 997-1020, (1985)
/2/ J.I.Pankove (Volume Editor), Semiconductors and semimetals, Volume 21, Hydrogenated Amorphous Silicon, Part A-C, (1984)
/3/ D.Adler, Solar Cells 21, 439-448, (1987)
/4/ K.Morigaki, J. of Non-Crystalline Solids, vol. 77&78, 583-591, (1985)
/5/ P.Viscor, Journal of Non-Crystalline Solids, vol. 77&78, 37-46, (1985)
/6/ C.R.Wronski, B.Abeles, T.Tiedje and G.D.Cody, Solid State Communications, Vol.44,No.10,pp.1423-1426,(1982)
/7/ R.A.Street, J. of Non-Crystalline Solids, vol. 77&78, 1-16, (1985)
/8/ S.Selberherr, Analysis and Simulation of Semiconductor Devices, Springer-Verlag Wien New York,(1984)

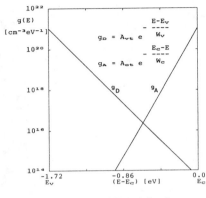

$A_{ot}=1.05980.10^{21}$ $(eV)^{-1}cm^{-3}$
$W_c=0.053eV$

$A_{vt}=9,54097.10^{20}$ $(eV)^{-1}cm^{-3}$
$W_v=0.088eV$

Fig. 1. Exponential distribution of gap states in a-Si:H

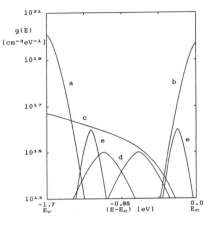

Fig. 2. Improved model of gap states in a-Si:H
a) tail of donorlike states
b) tail of acceptorlike states
c) donorlike states of strongly deformed bonds
d) defects with positive correlation energy
e) defects with negative correlation energy

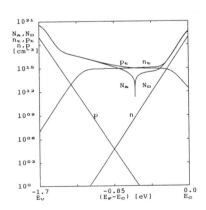

Fig. 3. Dependence of densities N_A, N_D, n, p, n_t, p_t on Fermi level in thermal equil.

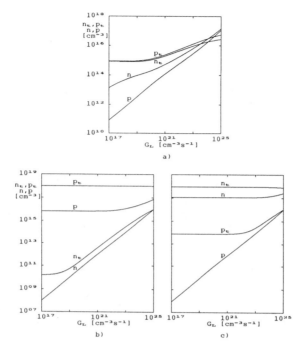

Fig. 4. Concentrations of free and trapped charge carriers as function of light generations
a) in an undoped a-Si:H
b) p-type doped with $N_A = 10^{18} cm^{-3}$
c) n-type doped with $N_D = 10^{18} cm^{-3}$

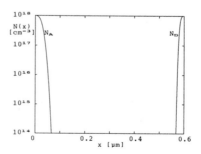

Fig. 5. Spatial impurity distribution in the examined p-i-n a-Si:H solar cell

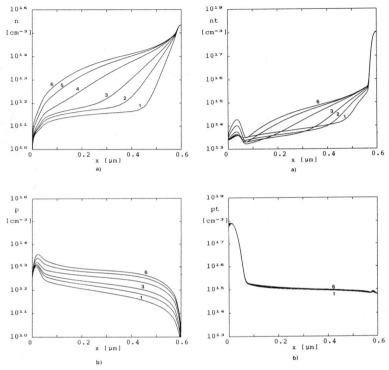

Fig. 6. Free charge carrier profiles for different applied bias voltages
a) free electrons
b) free holes

Fig. 7. Trapped charge carrier profiles for different applied bias voltages
a) trapped electrons
b) trapped holes

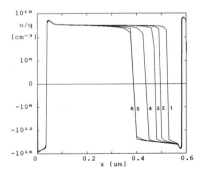

Fig. 8. Profiles of space-charge in illuminated a-Si:H p-i-n structure for different bias voltages

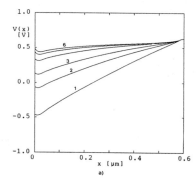

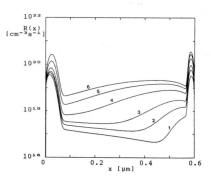

Fig. 10. Recombination rate profiles illuminated p-i-n structure

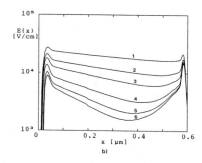

Fig. 9. Spatial profiles of
a) potential
b) electric field

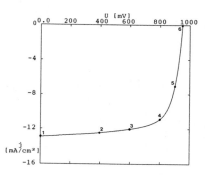

Fig. 11. Current-voltage characteristics of
illuminated (AM-1) p-i-n a-Si:H SC
μ_n = 20 cm²/Vs
μ_p = 1 cm²/Vs
$v_{th}\sigma_N$ = 10^{-11}cm³/sec
R = 100
U_o = 950mV
J_{sc} = 13mA/cm²
FF = .7
η = 8.7%

A NON-PHOTOVOLTAIC APPLICATION OF AMORPHOUS SILICON: ELECTROLUMINESCENT DISPLAY WITH PIXEL MEMORY.

by I. Solomon[*], P. Thioulouse[**], and M. Hallerdt[*][†].

The research on the semiconducting properties of amorphous silicon and related alloys is an example of how the results of basic research can be transferred rapidly from the laboratory to industrial applications. In hardly more than a decade after the first publication on the possibility of an efficient doping of the material[1], the industrial applications of amorphous silicon have become an important business, in the range of several hundreds of millions of dollars per year.

The most popular and well known application of amorphous silicon is inexpensive photovoltaic solar cells. But less known non-photovotaic applications (Table I) account, at least in the short and medium term, for a business with a turnover which is larger than that of the photovoltaic applications. For further information, we recommend a very recent review on the subject by LeComber[2].

Table I. Some non-photovoltaic applications of amorphous silicon and related alloys.

- Electrophotography, Xerography.
- Displays, flat TV screens.
- Image sensors, scanners, Vidicons, color sensors.
- Electronic memories, optical memories.
- High power devices, high voltage transistors.

In this article, we present such a non-photovoltaic application: the important improvement of the performance of electroluminescent displays, by the use of the transport properties of thin films of amorphous silicon and its alloys.

I. THE ELECTROLUMINESCENT DISPLAY

The structure of an electroluminescent display is shown in Fig. 1. The active layer (the light emitting layer ZnS: Mn) is sandwiched between two dielectric layers. Two perpendicular sets of electrodes ("lines" and "columns"), for the electrical excitation of the device, are situated on each side of the pile (Fig. 2). One of the two sets of electrodes has to be transparent in order for the light generated in the active layer to be visible to the user. In general, these transparent conducting electrodes are made of indium-tin oxide (ITO) deposited directly on the glass substrate.

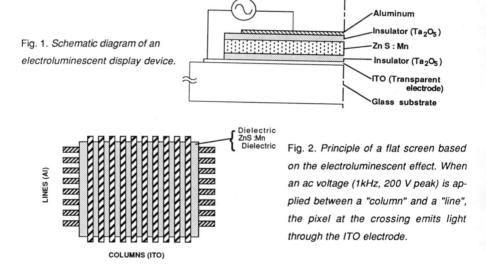

Fig. 1. *Schematic diagram of an electroluminescent display device.*

Fig. 2. *Principle of a flat screen based on the electroluminescent effect. When an ac voltage (1kHz, 200 V peak) is applied between a "column" and a "line", the pixel at the crossing emits light through the ITO electrode.*

The luminance-voltage curve of a typical electroluminescent (EL) cell is shown in Fig. 3. The strong non-linearity of this curve makes it possible to electrically address a large number of lines while keeping a high contrast. Typically, a commercial EL screen of 256 x 512 pixels has a luminance of 100 Cd/m^2 and an intrinsic contrast (ratio of luminance between an illuminated spot and a dark spot) greater than 30.

The EL display thus has some specific advantages. It uses an all-thin-film technique and it requires no active devices. So it is extremely simple to deposit without complicated masks or delicate alignment. It is possible to realize electroluminescent flat screens with high resolution , in practically any size or shape. Such screens have a high luminance and a large viewing angle (more than 160°) : The light intensity has a diagram close to that of Lambert's law ($\approx \cos \theta$). But above all, it is its high contrast in ambient light and at angles that makes it an attractive display.

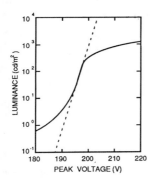

Fig. 3. *Typical response of the electroluminescent device of Fig. 1 (ac voltage at 1 kHz). The dashed line has a slope of 1 decade / 3 V.*

The price to pay for these advantages is two-fold. Firstly, the electrical excitation requires a high voltage, typically an ac voltage of 1 KHz and 200 Volt peak (Fig. 3). The total power consumption is rather high : 15 Watt for a 200 cm^2 screen. Secondly, the switching of high voltages is not compatible with standard integrated circuits. This makes the driving electronics of an EL display rather complex and very expensive. We will see in Section III that most of these difficulties can be overcome by combining the EL display with a photoconducting layer resulting in local memory ("Pixel memory").

II. PHYSICS OF AN ELECTROLUMINESCENT DEVICE

It is not possible to provide a complete discussion of the physics of EL in such a short article so we shall limit ourselves here to the basic mechanisms of the effect. For more detailed information, the reader is advised to consult a specialized review article, for example ref. 3, 4. Upon the application of a large electric field across the sandwich structure of Fig. 1, the different steps in the EL effect are the following (see Fig. 4) :

> 1 - Injection by tunneling of the electrons from the cathodic interface into the ZnS film;
> 2 - Impact excitation of the luminescent centers (Mn) by the electrons accelerated in the large applied electric field;
> 3 - Radiative de-excitation of the excited centers;
> 4 - Trapping of the electrons in the localized states of the other ZnS-Dielectric interface.

The phenomenon seems simple enough, but the actual fabrication of the Mn-doped, polycristalline film of ZnS involves some witchcraft: the EL efficiency of the material is quite sensitive to the deposition parameters, the post-thermal treatment and the amount of incorporated manganese.

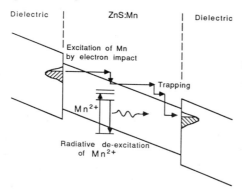

Fig. 4. *Principle of the electroluminescent effect : the Mn^{2+} centers, excited by electron impact, emit yellow light.*

III. AN ELECTROLUMINESCENT DISPLAY WITH PIXEL MEMORY

A very important improvement in the practical performance of an EL can be obtained with the "pixel memory" effect[5]. The schematic structure of a device incorporating this effect is given in Fig. 5: a photoconductive layer (PC) is deposited on the top of the oxide-ZnS-oxide sandwich. With this structure, the photoconductor is in series with the ZnS layer, so that the applied voltage can excite the active layer only through the photoconductor.

It is clear that such a device can have two stable states for a given applied voltage V_0:

1) - A dark state: the photoconductor is in the dark, and has a high impedance. The applied voltage V_0 is mostly across the PC layer and thus is not sufficient to drive the ZnS layer which remains in the dark.

2) - An illuminated state: In this state the photoconductor, illuminated by the electroluminescent layer, has a small impedance and the applied voltage V_0 is mostly applied to the ZnS sandwich. It is then sufficient to drive the active layer which thus remains illuminating.

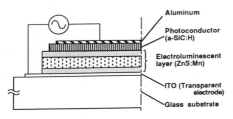

Fig. 5. *Electroluminescent display with pixel memory. The active EL structure of Fig. 1 is in series with a photoconducting layer. This "photoconductor" is a $n^+i\,n^+$ structure made of amorphous carbonated silicon (carbon concentration: 10%).*

To pass from state 1) to state 2) (dark to light), one has to apply a voltage larger than V_0 and, similarly, to pass from light to dark, it is necessary to decrease the applied voltage below V_0: the light versus applied voltage shows a hysteresis curve as shown in Fig. 6a. This provides a method to electrically activate ("write") or deactivate ("erase") the pixel (Fig. 6b): the application of a pulse on the top of the applied ac voltage illuminates the pixel (write) and the application of a negative pulse (decreasing the ac voltage) puts it in the dark state (erase). Once in one of the two states, the pixel remains in this state (dark or illuminated) as long as the sustaining ac voltage V_0 remains applied. It is interesting to remark that it is also possible to excite a dark pixel by a spot of light ("optical writing"), the EL screen then behaving as an electrical "slate-board" (Fig. 7).

This "pixel memory" effect, which is very easy to obtain (it requires the deposition of a thin film of a photoconducting material like a-Si:H, without any mask), has many advantages compared to the standard EL display:

- Since there is a local memory, the complete system requires no electronic memory. However, this is not much of an advantage since very large memories can be obtained now at low cost.

- More important is the decrease of the power consumption. In a standard EL screen, a large amount of power is spent in the modulation of the pixels of the picture[6]. With

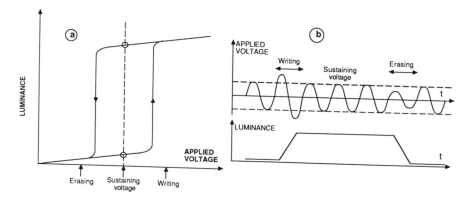

Fig. 6. *Schematic response of an electroluminescent display with pixel memory.*
a) Hysteresis curve. b) Electrical writing and erasing.

Fig. 7. *Optical writing of a picture on an electroluminescent display of 60x60 pixels with memory (surface: 2x2 cm^2). The message has been hand-written with a small light spot. The actual contrast is greater than 100.*

Table II. Electroluminescent display panel.
(Computed performance)

(256 x 512 pixels, 2 dm^2)	**Without Memory**	**With Memory**
Average luminance	45 cd/m^2	45 cd/m^2
Maximum consumption	15 W	3.5 W
Typical consumption	11 W	1.3 W
Peak current per line	40 mA	0.6 mA
Peak current per column	6 mA	70 µA

the pixel-memory effect, the modulation voltage is used only for the "refresh" of the information and the consumption decreases by almost 90% (see Table II).

- But the main advantage is that writing or erasing requires only the hysteresis voltage (Fig. 6), typically less than 20 Volts. This makes the electronics compatible with standard integrated circuits, providing a considerable decrease in the cost of the driving electronics and therefore of the complete system.

IV. THE SWITCHING PROPERTIES OF THE PHOTOCONDUCTOR

In practice, the photoconducting layer is a film of amorphous silicon (a-Si:H) or, more appropriately for this application[7], a film of carbonated silicon a-Si$_{1-x}$C$_x$:H [8]. We see in Fig. 8 that the device behaves as predicted. The intrinsic contrast (Luminance ON/ luminance OFF) is close to 3000, one of the highest values obtained for any type of display, the switching voltage is ± 15 Volts and the device is very stable in time[5].

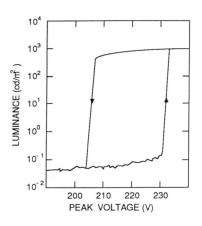

Fig. 8. *Actual hysteresis curve of an electroluminescent device with pixel memory. Operation in the dark with a 1 kHz sine-wave. We remark a very high intrinsic contrast (illuminated/dark ratio).*

If a large voltage is applied to the a-SiC:H layer, the space-charge limited current (SCLC) regime[9] is attained. However, the applied voltages vary rapidly, so the study of the static response of the system is not appropriate in this case. The amount of research on the dynamical behavior of thin films of amorphous silicon and alloys is very reduced compared to that on the static or photovoltaic properties of the material. In a pioneering work, den Boer et al.[10] have shown that the transient transport in films of amorphous silicon can have interesting switching properties. More recently, we have made a systematic study[11] of the response of this type of materials to the switching-ON and switching-OFF of large voltages. We give here a summary of the results relevant to the present application of EL display with memory.

1°) Switching properties in the dark.

The "photoconductor" in the device shown in Fig. 5 is an n^+in^+ structure, with a total thickness of about 1μm. The application of a large ac voltage in the dark produces an electron injection through both sides of the structure, thus maintaining a large space charge which results in a voltage barrier V_s across the film (Fig. 9). Depending upon the amount of injected electrons, amount that can be controlled by the amplitude of the applied voltage V_a, the height of the barrier can be varied: this is a definite advantage over a static barrier which has a fixed height by construction. As calculated in the Appendix, the value of V_s can be controlled up to a maximum value that can be very large, typically more than 50 volts. This is much more than what can be obtained with a static device: with diodes, the same voltage would require some 70 diodes in series !

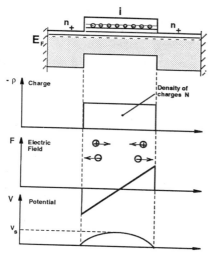

Fig. 9. *The injected electrons from the n^+ layers populate the electronic states in the gap, producing a space charge in the film. The value of the potential barrier V_s resulting from this space charge is given by Eq. (1) in the Appendix.*

This switching voltage V_S is very sharp; below that voltage, the current is practically zero, and above, the current increases very fast. The switching ratio can be as high as 10^4, which makes it it suitable for applications to other types of displays, e.g. liquid-cristal device[10].

2°) The effect of light.

The irradiation of the film with visible light does not produce a simple photoconducting effect as presented in Section III for the sake of simplicity. In fact, the switching properties are dominated by the quasi-equilibrium space charge in the film, and the effect of the light is to decrease the space charge and thus the voltage V_S across the film.

Light irradiation produces electron-hole pairs in the film (Fig. 10). Due to the very large internal electric field in the material resulting from the space charge, the created electrons will be swept out of the film whereas the created holes will be attracted inside the film where they will eventually recombine with trapped electrons in the space charge. This transport of the carriers is very fast: for a voltage V_S of 50 volts across a film of 1µm and for a mobility of the photocreated carriers of 1 cm^2 V^{-1}s^{-1}, the transit time of the carriers is 2.10^{-10} s, much faster than the recombination. The net effect (negative charges out, positive charges in) is a decrease of the negative space charge and therefore a decrease of the barrier V_S. The calculation in the Appendix shows that for 3.10^{15} photons s^{-1}cm^{-2} absorbed in the film (a light of about 1000 Lux), the barrier V_S decreases by about 27 volts. Such a decrease of the voltage across the a-SiC:H film increases the voltage across the active ZnS layer by the same amount, thus producing the "pixel memory" effect.

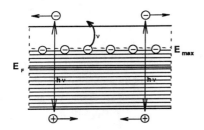

Fig. 10. *The gap states of the amorphous silicon-carbon films can be populated up to an energy level E_{max} given by Eq. (2). Under illumination, the photo-created electrons are swept out into the sides and the holes will bring then a net positive charge, thus decreasing the barrier voltage V_S.*

V. CONCLUSIONS

We have shown that the space charge distribution injected in the intrinsic layer of a n^+in^+ structure made of carbonated amorphous silicon a-$Si_{1-x}C_x$:H can be used as a dynamical switch to drive an electroluminescent device. The switching voltage of the film can be tuned to any value up to about 60 volts by controlling the amount of injected electrons. The height of the barrier due to the injected space charge decreases under illumination, and we have used this effect to make a device with memory. This "pixel memory" effect brings an important improvement in electroluminescent displays. The benefits are an improved contrast, and lower switching voltages compatible with standard integrated-circuit electronics. All this is obtained with very simple thin-film technology, by-passing the complication of expensive masking levels or difficult alignment procedures. More generally, the very sharp switching properties of a-Si:H thin films can be used to drive all kind of flat-screen displays.

APPENDIX

We apply an ac voltage across the n^+in^+ structure, thus injecting electrons through both side of the intrinsic layer. It is then a good approximation to consider the space charge distribution across the film as constant. For such a distribution, a simple integration gives the equation between the trapped charges and the voltage V_s across the layer[11] (Fig. 9):

$$V_s = \frac{1}{2} \frac{e}{\varepsilon \varepsilon_0} N L^2 \qquad (1)$$

where N is the density of the space charge in the film of thickness L. It is possible to vary the height of the barrier by injecting and trapping in the gap different amount of electrons. In particular, if we apply periodically a voltage V_a (e.g. an ac voltage of amplitude V_a) and if the relaxation of the space charge is very slow compared to the period of the applied voltage, one has simply $V_s = V_a$. However, there is a limitation to the amount of charges that can be trapped in the gap. The injected electrons will populate the electron states in the gap up to an energy level E_{max} (Fig. 10): at this energy and above, the electrons are thermally excited back into the conduction band at a rate which becomes equal to or faster than the rate of injection, i.e. the period of the applied voltage. The value of E_{max} is thus given approximately by:

$$\nu = \nu_0 \, e^{-\frac{E_c - E_{max}}{kT}} \qquad (2)$$

where ν is the frequency of the applied voltage, ν_0 is the "attempt-to-escape" frequency (typically 10^{11} s^{-1}) and E_c is the energy of the conduction band edge. For $\nu = 1$ kHz, this gives a value $E_c - E_{max}$ of about 0.48 eV.

The barrier V_s due to this maximum space charge in the film can be very high. In an actual device, the photoconductor is a film of a-Si$_{1-x}$C$_x$:H with the following properties:

- Carbon concentration $x = 0.1$
- Activation energy $E_c - E_F = 0.85$ eV
- Density of states in the gap   (average) $\approx 2.10^{17}$ eV^{-1} cm^{-3}
- Thickness $L = 1\mu m$

This gives a density of charge in the gap $N = 7.4 \; 10^{16}$ electrons/cm^3, and from Eq.(1), a typical voltage $V_s = 64$ volt.

When the film is exposed to light, the absorbed photons will create electron-hole pairs. The electrons will be swept out into the sides of the film and the holes will then bring a net positive charge, thus decreasing the barrier voltage V_s. For 3.10^{15} absorbed photons per second per cm^2 (light irradiation of about 1000 Lux) in a film of 1μm, the space-charge density is reduced at a rate $dN/dt = 3.10^{19}$ electrons s^{-1} cm^{-3}. If the rate of the space-charge build-up is, according to Eq.(2), $\nu = 10^3$ s^{-1}, we have then a decrease of $\Delta N = 3.10^{16}$ electrons/cm^3 in the space charge which corresponds, from Eq.(1), to a decrease of the barrier $\Delta V_s = 27$ volt.

References

* Laboratoire de physique de la Matière Condensée, Ecole Polytechnique, 91128 Palaiseau Cedex, France.

** Centre National d'Etudes des Télécommunications, 192 Avenue Henri Ravera, 92220 Bagneux, France.

† Permanent address: SOLEMS S.A., 3 rue Léon Blum, Z.I. Les Glaises, 91120 Palaiseau, France.

1. W.E.Spear, and P.G.LeComber, Phil. Mag. **33**, 935, (1976).
2. P.G.LeComber, *Proceedings of the 8th European Photovoltaic Solar Energy Conference, Florence 1988,* Ed. by I.Solomon, B.Equer, and P.Elm, Kluwer Academic Publications, Dordrecht, The Netherlands (1988) p.1229.
3. R.Mach, and G.O.Müller, Phys. Stat. Solidi (a), **69**, 11 (1982).
4. Y.Hamakawa, R.Fukao, and H.Fujikawa, OPTOELECTRONICS-Devices and Technologies, **3**, 31 (1988).
5. P.Thioulouse, C.Gonzalez, and I.Solomon, Appl. Phys. Lett., **50**, 17 (1987).
6. P.Thioulouse, thesis, Doctorat ENST, Paris (1987) (unpublished).
7. P.Thioulouse, and I.Solomon, EUR Patent n° 86 14715 (1986).
8. I.Solomon, M.P.Schmidt, and H.Tran-Quoc, Phys. Rev. B **38**, Nov. 15, (1988).
9. I.Solomon, R.Benferhat, and H.Tran-Quoc, Phys. Rev. B **30**, 3422 (1984).
10. W.den Boer, and A.F.T.Pop, Solid State Comm. **45**, 881 (1983).
11. M.Hallerdt, thesis, Doctorat Université Paris-Sud, Paris (1988) (unpublished).

Application of a-Si Diodes to Liquid Crystal Display

K.Urabe,M.Kamiyama,E.Tanabe,and H.Sakai
Fuji Electric Corporate Research and Development,Ltd.
2-2-1 Nagasaka,Yokosuka,Kanagawa,Japan

Introduction
     Liquid crystal displays (LCDs) have been widely used in watches and calculators. In recent years LCDs have been applied to devices which require large area and high resolution displays;e.g.,computer terminals and flat panel televisions. Such high-resolution LCDs are addressed by multiplexing method which uses only the threshold characteristics of liquid crystal. However, the contrast ratio and viewing angle are not sufficient for such applications because liquid crystal is lacking in electro-optic threshold sharpness required for multiplexing.To obtain excellent image quality,an active matrix display is used. In the active matrix display, an individual switching element is provided for each picture element to drive the liquid crystal in an effectively static mode. The active matrix elements are used for switching the information stored in each picture element. In the writing mode, the resistance of the active element must be low, and in the storage mode, the resistance of the element must be high enough to hold the written information for an entire frame time.
     Two types of the switching elements are used in the active matrix display. One is two-terminal element having nonlinear current-voltage characteristics (e.g.,diode).The other is three-terminal element having an independent control gate (e.g.,transistor).Thin film diodes (TFDs) and thin film transistors (TFTs) using hydrogenated amorphous silicon (a-Si) have been investigated for application to flat panel displays.
     The advantage to use the a-Si is as follows;
(1)a-Si films can be deposited on a large area substrate by plasma CVD.
(2)A glass substrate can be used for a-Si devices,because the plasma CVD is a low temperature process.
(3)Well developed photolithographic techniques are applicable to

obtain fine patterned switching elements.

Figure 1 shows the fabrication process of the a-Si TFT. The main processing steps are as follows:

(1) An indium-tin-oxide(ITO) and a metal layer are formed on a glass substrate and patterned.

(2) A dielectric layer(Si_3N_4) and an a-Si layer are deposited by plasma CVD and patterned.

(3) A highly phosphorous-doped n^+ type a-Si layer is deposited on top of the undoped a-Si layer to form the source-drain contact.

(4) A metal layer is formed for source and drain electrodes.

In the fabrication process of the a-Si TFT, six or seven kinds of masks are required. The process is too complicated and reproducibility is not well enough. In addition, the presence of crossing lines on the same substrate increases the probability of creation of defects. Another problem is that it is difficult to obtain a high quality dielectric for the gate which affects a threshold stability.

To overcome the problems in the process of the a-Si TFT, we have developed the a-Si TFD. Because the fabrication process of

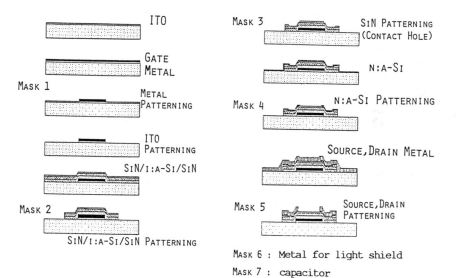

Figure 1. Fabrication process of a-Si TFT

the a-Si TFD is simpler as compared with that of the TFT,the a-Si TFD is suitable for fabricating large area displays.The LCD panel with the a-Si TFD which we have developed has 240 x 480 picture elements on a 5.5 inch diagonal display area. In this paper, we will present and discuss the structure and the characteristics of the a-Si TFD.

Structure and Drive Method of a-Si TFD-LCD

Figure 2 shows the structure of the active matrix LCD addressed by the a-Si TFD. Figure 3 shows its equivalent circuit.The a-Si TFD and the liquid crystal element in each pixel are connected in series between a scanning line and a data line.To drive the LCD with an AC signal, the a-Si TFD element consists of a pair of diodes connected in parallel with opposite polarities,i.e., in a ring configuration. Therefore, it is referred to as a diode ring. The current-voltage characteristics of the diode ring consist of the forward bias diode current.We use the forward bias diode current for driving liquid crystal elements.

Figure 4 shows the waveforms for driving the LCD addressed

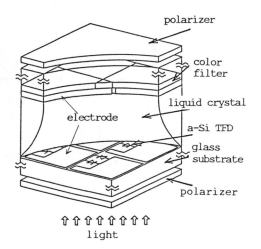

Figure 2. Structure of active matrix LCD

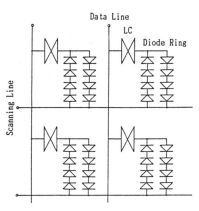

Figure 3. Equivalent circuit of active matrix LCD

by the a-Si TFD. The scanning signals consist of the writing phase(60 μ sec) and the holding phase(17msec). During the writing and the holding phase, the writing voltage (Va= $\pm$ 6V) and the holding voltage(Vb= $\pm$ 2V) are applied, respectively. The data signal has two levels(Vc=$\pm$ 1V) according to the ON and OFF state of LCD. Levels of the range from $-$Vc to $+$Vc correpond to a gray scale. During the writing phase, the diode current (i.e., the ON current) must be large enough to store the charge in liquid crystal elements and the OFF current must be small enough to hold the stored charge during the holding phase, that is, it is required that the ON current is larger than 1×10^{-7}A and the OFF current is less than 1×10^{-11}A.

Figure 5 shows the current-voltage characteristics of the a-Si TFD. The threshold voltage of a single diode is 0.5V. On the other hand, the threshold voltage of liquid crystal is between 1.5V and 2.0V. Thus it is difficult to hold the charge effectively by a single diode. Therefore, we used a four-series-connected diode to increase the threshold voltage.

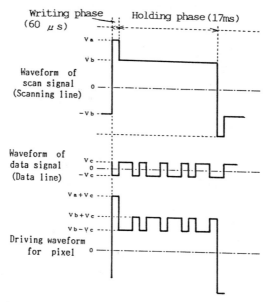

Figure 4. Waveforms for driving a-Si TFD-LCD

Fabrication process of a-Si TFD

The structure of a pixel is shown in Figure 6. Figure 6(a) shows a schematic top view of the pixel. Figure 6(b) shows the cross sectional view of the four-series-connected TFD array along A-A' of Figure 6(a). The ITO layer acts as pixel electrodes. The TFD has three a-Si layers to form a p-i-n junction. The a-Si layer is sandwiched between two chromium (Cr) layers. They prevent inter-diffusion between the ITO and the a-Si layers and shield the a-Si layers from light incidence. An insulating layer is composed of silicon nitride film. An aluminum (Al) layer forms the wiring to connect the Cr layer on the a-Si layer to the scanning line electrode.

Figure 7 shows the fabrication process of the a-Si TFD. First, a 1000Å-thick ITO layer is deposited as a transparent conductive thin film on a glass substrate. Next, a 1000Å-thick Cr layer is deposited by sputtering. An a-Si p-i-n layer is deposited by means of rf glow discharge decomposition of silane (SiH_4). Deposition conditions are a substrate temperature of 190 ℃, a chamber pressure of 1 Torr, and an rf power of 50W.

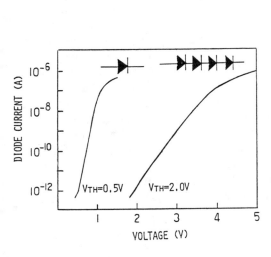

Figure 5. Current-voltage characteristics of TFD

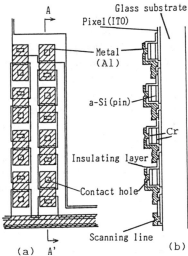

Figure 6. Structure of a pixel

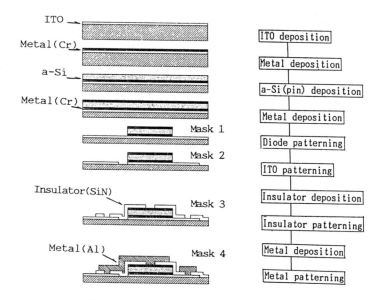

Figure 7. Fabrication process of a-Si TFD

The a-Si layers are fabricated by depositing 400 Å -thick p-type layer($B_2H_6/SiH_4 \sim 1\times10^{-2}$), and then subsequently depositing 3000 Å -thick undoped layer. Finally n-type layer($PH_3/SiH_4 \sim 1\times10^{-2}$) of 400 Å thickness is deposited. A Cr layer is then deposited by sputtering on the a-Si layers. The Cr/a-Si/Cr layers are patterned by photolithographic technique. The Cr layer is patterned by dry etching using a gas mixture of CCl_4 and O_2, and the a-Si layer is etched by using CF_4 and O_2. The area of the patterned a-Si layer is 10 X 10 μ m². Next, the ITO layer is patterned for the pixel electrodes, and the silicon nitride film is deposited by plasma CVD and patterned. Finally the a 1 μ m-thick Al layer is formed and patterned to connect the a-Si layer and the scanning line electrode. In the process above, only four kinds of masks are required.

Characteristics of a-Si TFD

To obtain a uniform picture all over the display area, all of TFDs on the glass substrate are needed to have uniform current-voltage characteristics. In order to examine uniformity of current-voltage curve, we picked out 2560 TFDs from 115,200 pixels and measured their distribution of current biased at 2V.Figure 8 shows the current distribution of the a-Si TFD substrates. Figure 8(a) shows an example of good sample with uniform current distributions,but some TFD substrates do not show flat distributions as shown in Figure 8(b).

Next,we picked out at random 40 TFDs from one TFD substrate and measured their current-voltage characteristics. Figure 9 shows the forward current-voltage curves of TFDs consisting of four-series-connected diodes. The curves of 40 TFDs are shown together in the same figure.In the TFD substrate shown in Figure 9(a),the curves are almost the same and the OFF current,the current at the applied voltages lower than 2V, is less than 1×10^{-11}A. On the contrary,the curves of the TFD in Figure 9(b) scatter. In addition,the OFF current is larger than that of the TFD substrate in Figure 9(a). Because the off-resistance of the TFD is too low to hold the stored charge in liquid crystal elements,pixels do not show the ON state.

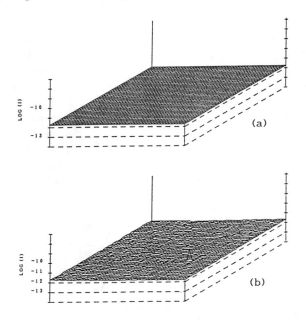

Figure 8. Current distribution of TFD substrate

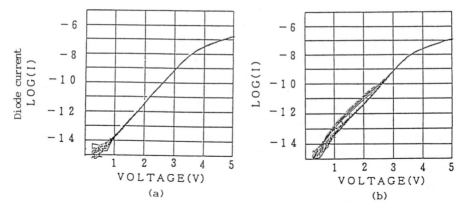

Figure 9. Forward current voltage characteristics of TFD

We observed patterned diodes by scanning electron microscope (SEM) at magnifications of 60000. Figure 10 shows edge profiles of Cr/a-Si/Cr layers. Figure 10(a) shows the profile of the patterned diode with uniform current-voltage curves,and Figure 10(b) shows that of the diode which has not uniform curves.The profile in Figure 10(a) is nearly vertical.In contrast,the profile in Figure 10(b) has a taper and an undercut at the upper Cr layer edge. We can remove easily contaminations on the vertical edges.However,it is difficult to clean the undercut parts sufficiently.When the contaminations remain on the edges of the diodes after cleaning,leakage current is likely to appear along the edges.It is suggested that the increase in the OFF-current is due to the leakage current and the vertical edge profiles are desirable to decrease the leakage current.

The edge profiles are affected by the etching process conditions(gas pressure and rf power).We tried to optimize the conditions to form vertical edges.When the gas pressure was higher than 5 Pa and the rf power was lower than 1kW, the diodes had taper profiles.On the other hand,when gas pressure was 3 ∼ 5 Pa and rf power was 1∼1.5 kW,the profiles were vertical.At a lower pressure and a higher power,the etching in the vertical direction is greater than in the lateral direction.Thus we have optimized the dry etching conditions and obtained the TFDs with uniform characteristics.

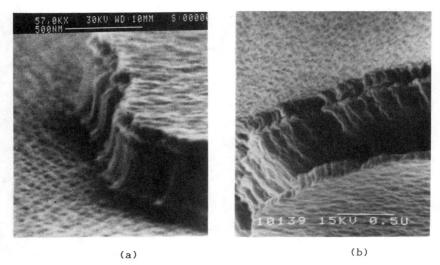

(a)   (b)

Figure 10. SEM photograph of patterned diode

Display Characteristics

Figure 11 shows light transmittance-voltage characteristics of the LCD with a-Si TFD. To obtain the ON and OFF state of the LCD, the writing voltage(Va) requires from 5 to 7V when the holding voltage(Vb) is 2V. Figure 12 shows the contrast ratio of the LCD as a function of the viewing angle. The cotrast ratio is larger than 30 for the range of from -20 to $+20$ degrees at the horizontal and vertical viewing angles respectively.

Conclusions

We have developed the a-Si TFD for the 5.5 inch sized LCD. By optimizing the dry etching condition, we have obtained uniform current-voltage characteristics of the TFDs. The a-Si TFD-LCD is promising for obtaining large area displays, because it has the advantages of high display qualties and simple fabrication process.

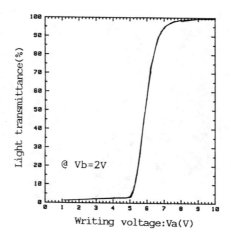

Figure 11. Transmittance-voltage characteristics of a-Si TFD-LCD

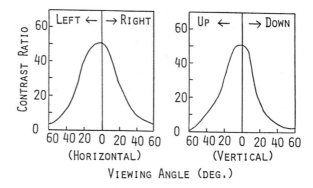

Figure 12. Contrast ratio of a-Si TFD-LCD

References
1. P.G.LeComber et al.,Electronics Letters,15,179(1979)
2. S.Togashi et al.,Proc.SID.,vol.26/1,p.9(1985)
3. Z.Yaniv et al.,MRS Symp.Proc.vol.70,p.625(1986)
4. H.Sakai et al.,MRS Symp.Proc.vol.118,p.375(1988)

POSSIBLE APPLICATIONS OF AMORPHOUS SILICON IN NUCLEAR AND X-RAY PHYSICS

C.Manfredotti

Experimental Physics Dept., University of Torino, Torino (Italy).
and
Sezione INFN di Torino (Italy)

ABSTRACT
Recent applications of amorphous silicon to α-particles, X-rays and ß particles detection are reviewed. Charged particles spectrometry and X-rays detection for industrial, medicine and synchrotron radiation experiment application is proved to be a real possibility with the presently available material or with an adaptation of market area image sensors. Detection of minimum ionizing particles and applications to tracking and to calorimetry in the next high energy collider experiments may be achieved with stack arrangements or with pixel geometry. If the material quality will improve, many fields of application will be opened in all the branches of nuclear and X-ray physics.

1 - INTRODUCTION

Amorphous silicon has been considered so far only for photovoltaic and optoelectronics applications, and some other applications in a typical electronics field have been proposed[1,2]. At a first glance, a-Si:H has indeed a too large defect density, and too low trapping times and carriers mobilities, to be considered for nuclear particle detection. In fact, depletion layer thicknesses, collection lengths and signal risetimes may not be adequate for some applications. However, the cheap production of a-Si:H films over large areas may offer exciting possibilities which are precluded to c-Si, both for technical reasons and for price. Therefore, the possible applications of a-Si:H in nuclear physics must be seriously considered and experimentally investigated, before to draw any final conclusion. In fact, α-particle[3], protons[4] and X-ray detection[5] has been successfully carried out and, moreover, optoelectronics components may find important applications in CAT scanners[6] and in synchrotron radiation detection[7]. Moreover, other exciting applications may be envisaged, particularly in high energy and heavy ion physics.

In the present paper, the more relevant achievements in nuclear particle detection will be shortly reviewed, considering separately nuclear physics and medicine and industrial applications (X-rays). Some suggestions will be also advanced and discussed.

2 - NUCLEAR PHYSICS APPLICATIONS: HIGH ENERGY EXPERIMENTS

High energy physics, going now towards TeV region of center of mass energy, demands for ever bigger and more expensive detection apparatuses, with a good granularity and with accurate vertex detection systems. A possible apparatus for SSC (Superconducting Super Collider), a multi-Tev USA hadron-hadron collider of the next generation, will consist of 500 K detection channels for a total cost of 500 M\$. The vertex detector, which is expected to be realized with silicon detectors, should contain $3.2 \cdot 10^6$ channels, each reading 1.2 m long strips of silicon and, quoting detectors at 200\$/cm^2 and an electronics at \$100/channel, should globally cost 600 M\$[8]. The microstrip silicon detectors represent a well established technology, which is, however, extremely espensive and still not satisfactory, particularly as far as radiation damage is concerned. Moreover, the problems associated with mounting and aligning such detectors, by contemporarily maintaining a spatial resolution and a precision for track reconstruction in the μm range, are absolutely not trivial. Of coarse, large area detectors are desired, but present silicon technology is limited to 6″ wafers. The huge number of electronics readout channels poses also severe space and heat dispersion problems, because pramplifiers, sample-and-hold circuits, etc. must be closely connected to detectors and the room around the vertex detectors is generally occupied by tracking chambers, calorimeters, muon detectors and so on (see Fig. 1). Moreover, in order

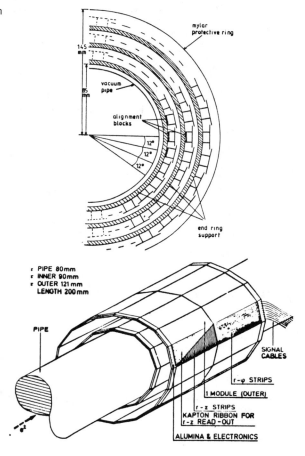

Fig. 1 - Arrangement of vertex detector in a typical detector for a collider experiment.

to be able to detect tracks in two dimensions, by avoiding errors due to multiple tracks, the microstrips detectors must be assembled and aligned along the x, y, and u axis, this last 45° rotated with respect to x (Fig. 2). For this reason pixel arrays have been suggested (Fig. 3), with a random access readout, which could meet the requirements for time and space resolution and could compete favourably in these respects with CCD and strip detectors. Radiation hardness should be also intrinsically higher, both because of the much lower dark currents (in the range of pA) and capacitance (below 0.1 pF).

Summarizing, the a-Si approach should present the following advantages:
1) reduced cost;
2) possibility of covering large areas, avoiding this way mounting and aligning problems:
3) possibility of vertical integration for pixel devices, also by using a-Si FETS circuitry, with a power consumption comparable to CMOS circuitry;
4) possibility of realizing and using a-Si CCDS, with probably a much larger radiation resistance;
5) possibility of realizing complex geometries by using photolitography directly to define deposition patterns;
6) possibility of automatization of the full production process, like for photovoltaic cells;

Let us now discuss briefly, but separately, the possible applications of a-Si in the various high energy particles detector types.

2.1. - Microstrip Detectors (MSD)

Let us assume that we are able to almost deplete 20 µm of thickness: according to calculations, this would require a 10^{14} cm^{-3} charge defect

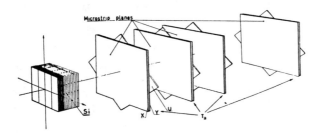

Fig. 2. Vertex detector for EG17 Fermilab experiment

Fig.3. Proposed pixel array connected by indium bump bonding to a bucket brigade readout

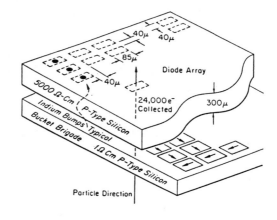

density material, wich is not too far from what is presently available. A strip 25 μm x 10 cm should have about 12 pF with a noise of no more than 10 keV, according literature[9]. Since a minimum ionizing particle loses most probably 26 keV/100 μm, S/N should be 0.6 approximately. Therefore, in this case, detectors should be stacked on top of each other, in order to increase S/N. A possibility is to connect each layer to a separate charge-sensitive preamplifier with a comparable input capacitance: for a six layers stack, S/N should be larger than 12, which is not a bad standard for the present MSDs.

If the non-capacitive part of noise is really a dark current shot noise and therefore it is proportional to the square root of the surface area of the detector, it can be further reduced to 4 keV/strip. If so, S/N > 1 should be reached without stacks and, with a low noise electronics, S/N >> 1 may be obtained even without a parallel connection of charge-sensitive preamplifiers. As a matter of fact, a two-layer stacked detector displayed no noise increase even if the area was doubled [3]. In the case of no noise increase, S/N = 10 for an 8 detector stack. For a totally depleted detector, 20 μm thick, at $3 \cdot 10^5$ V/cm average electrical field, pulse risetime should be 3 ns for electrons, but about 0.7 μs for holes. Therefore, hole collection must be excluded. If really the high field region is limited only to 4 μm for a 10^{15} cm^{-3} material, about 10 μm of the detector should have on electrical field below 10^2 V/cm and therefore only diffusion should occur, with total collection times of the order of 10 μs. This is really a drawback of a-Si detectors, which presently show a pulse saturation for α-particles at integrating times of the order of 1 μs (see Fig. 4) even for much thinner sensitive depths.

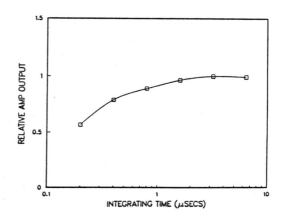

Fig. 4. Output signal for α-particles as a function of the peaking time from a charge-sensitive pre-amplifier attached to a 2μm a-Si H (120 pF) detector

2.2. - Pixel Detectors

Of coarse, a pixel of 40x80 μm^2 should have a very low capacity and, correspondingly, a very low electronics noise. Also the shot noise should be comparatively reduced. Let us assume, therefore, on more safe grounds, that minimum achievable noise is 1 Kev. In this case, S/N = 5 for a single detector and S/N = 10 for a double stacked pixel detector. Timing problems should not occur in this case, if another fast detector is used for triggering: reading times of the order of 1 μs for interesting events are allowed.

Therefore, an a-Si pixel detector separately triggered could be a real possibility.

2.3. - Calorimeters

A calorimeter is devoted to obtain the energy deposited by electromagnetic or hadronic showers and to calculate the direction of incoming particle which produced the shower. To this purpose, scintillators or silicon detectors, which measure position and energy, are generally mixed up with absorbers, which are high-Z materials for electromagnetic calorimetry and low-interaction length materials for hadronic calorimetry (Fig. 5).

Fig. 5 - Exploded view of one module of WA78 electromagnetic calorimeter

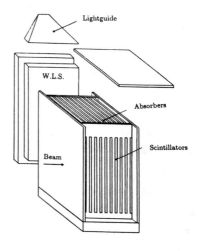

Scintillators must be coupled to PMTs, which, very often, cannot be used because of the presence of magnetic fields or space requirements. Therefore, silicon photodiodes have been suggested to this purpose, particularly high-gain, avalanche type (APDs). Another possibility, which has been advanced and largely investigated in recent years, is to use silicon detectors. The SICAPO electromagnetic colorimeter [10] consists of 24 radiation lengths of tungsten (12 cm), in which a silicon detector was located every two radiation lengths. For this kind of application, since the mean energy detected by the colorimeter is 5

MeV per GeV of incoming electron, each detector "reads" a fraction of MeV and, consequently, even underdepleted silicon detectors can be used. A longitudinal shower development in Si/W sampling colorimeter, produced by a 4 GeV electron, is shown in Fig. 6. Even 40 μm depletion

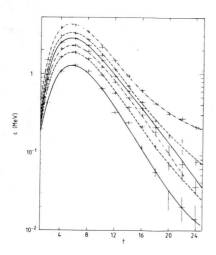

Fig. 6 - The longitudinal shower development in Si/W SICAPO sampling calorimeter for Si depletion depths of 40, 70, 100, 125, 150 and 200 μm. Incoming electron energy is 4 GeV.

widths may be used, with a good shower energy resolution. In hadronic calorimetry, large area detectors are required but, since the active volume can be small, relatively low resistivity, not fully depleted devices can be used. A proposed Si/U calorimeter will consist of 130 silicon planes, each with a mosaic of 18 trapezoidal 28 cm^2 silicon detectors, for a total number of 2340 detectors and a total area of 6.5 m^2. Total cost is in the region of several M$. Since the capacitance of each detector is large (about 150 pF for a 400 μm depletion) full depletion may be convenient in order to minimize the electronic noise. However, since these detectors are self-tested by α particles, partial depletion can be useful in order to correct for signal loss due to radiation damage. Physics of hadronic colorimetry is still not well understood and rely practically only on Monte Carlo calculations. Electrons, and recoil protons are the more important contributors to deposited energy, while minimum ionizing particles are less important. Protons in MeV range have been detected by a-Si detectors[9]: consequently, an application of a-Si detectors both to electromagnetic and hadronic colorimetry may be really affordable. Clearly, the calorimeters should be entirely redesigned in order to take into account the different detector properties of amorphous silicon, which, on the other hand, may be deposited directly on the absorbers and connected externally to electronics. Large areas could originate capacitance and, consequently, pulse rise-time and noise problems. In particular, this last one could amount to several hundred keV: more restricted areas should be therefore adopted if calorimeter geometry does not allow enough energy to be deposited in each detector.

Other electromagnetic colorimeters make use of BGO scintillators coupled directly to photodiodes. For instance, in L3 experiment 12.000 BGO crystals are used, $24 \times 2 \times 3$ cm^3 of volume, each of them viewed by two pin PDs of 1.5 cm^2 area. Amorphous silicon has a better spectral response with respect to c-silicon for practically all scintillators, but for large area, as in these cases, capacitance is much larger (for 1 µm thickness, we get for a-Si p-i-n PDs 10^4 pF/cm^2 as compared to 75 pF/cm^2). Certainly, a vertically integrated image sensor, which is directly compled to an a-Si FET, should be more adequate, even if also in this case the signal rise time will be too long. Effectively, reading times of 10 µs have been measured in these sensing elements, essentially because of the long lifetime in a-Si photoconductors.

Noise too needs some consideration: in Hamanatsu PDs, for 1 µs pulse shaping time constant, we have an equivalent noise charge of 1000 e^-, corresponding to about 3 MeV of incident photon energy and to about 6 KeV a-Si equivalent. From present noise data, we may expect about 300 keV/cm^2 from 1μm thick photodiodes. Particular low capacitance photodiodes must be developped for this kind of application.

Another application is also possible: a-SiC or a-SiN alloys can be directly deposited on scintillators and used as optical waveguides to collect light and to increment S/N value. The a-Si photodiode may be easily integrated in the film itself at the scintillator boundary. Of coarse, also in these cases, particular scintillator-photodiodes assemblies must be produced and tested, if the market seems to be large enough.

3. - NUCLEAR AND HEAVY ION PHYSICS APPLICATIONS

In this case, since low energy particles, such as α-particles from natural radioactive sources, and heavy ions possess a large value of specific ionization, giving large signals, results prove that possible applications may be advanced on a more realistic basis. Kaplan and Perez-Mendez data obtained at Lawrence Radiation Laboratory are the only ones quoted in literature, that allow some considerations and predictions. Let us summarize these results. Firstly, α-particles were detected by 2 μm thick a-Si pin diode biased at 90 V. This obviously means an electrical field of about 5×10^5 V/cm, with an electron average drift velocity of 10^6 cm s^{-1}, which is not so far from electrons maximum drift velocity in c-Si. Pulse height corresponded to only 60 keV deposited energy (see Fig. 7), which indicated clearly the presence of some loss mechanisms, since for 2μm sensitive depth, the signal should correspond to 260 KeV. A former mechanism is likely due to the different value of pair creation energy in a-Si, because of the larger

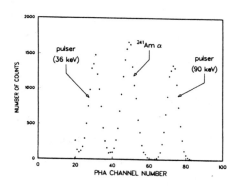

Fig. 7–PHA spectrum (Ref. 11), obtained by a 2μm a-Si: pin diode based at 90 V. The energy equivalents of the pulser calibration peaks were determined both by a direct comparison with a full energy alpha pulse in a c-Si detector and by calculation, from the value of the test capacitor.

energy gap. More precise measurements indicate a pair creation energy of 6.3 eV, compared to 3.6 eV for c-Si (see Fig. 8). A second, more subtle

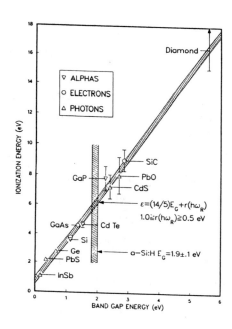

Fig.8–Radiation ionization energies in (eV/e-h pair) versus band gap energy for a number of semiconductors materials and for a-Si

argument, was due to the large degree of charge recombination produced in the high density of electron hole pairs of the α-track. When the detector was rotated of 60° with respect to the direction of incident α's, there was an almost 4-fold increase in signal size, as compared to a 2-fold one expected from geometry considerations. Data are reported in Fig. 9 and compared with protons (1 and 2 Mev) data both for different orientations and for different bias. This figure suggests that recombination is still a problem and that depletion is not enough to allow a complete collection. Fortunately, an extrapolation to minimum

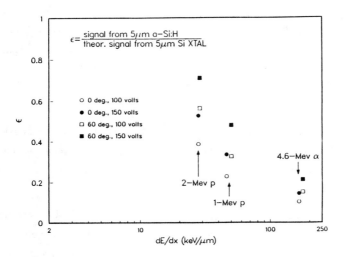

Fig. 9 - Charge collection efficiency vs. dE/dx for a 5μm pin diode

ionizing particles (0.26 KeV/μ in silicon) indicates clearly that complete collection is possible. This datum is in agreement with laser induced ionization data, which show full charge collection from diodes 4 to 7 μm thick and with E fields as low as 5×10^2 Vcm^{-1} (12).
Another further step was represented by the realization of a double-

layer detector, a nip diode deposited on top of a pin diode, which were connected in parallel. The signal, captured from a middle electrode, was practically doubled (see Fig. 10). The noise from these detectors was not large (about 3000 e, which means about 11 keV) and it is due to

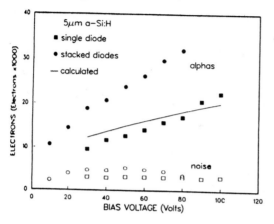

Fig.10 - Comparison of signal and noise from a single 5 μm pin detector and a stacked nip-pin detector. The noise reamains unchanged and the signal is doubled

the capacitance (see Fig. 11) with a lower contribution probably from dark current shot noise. Moreover, it seems not to depend on detector thickness and on the applied bios.

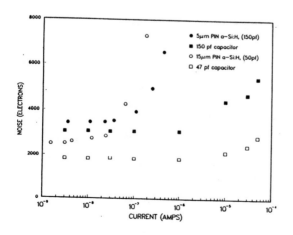

Fig. 11-Equivalent amplifier input noise from 5 μm (150 pF) and from 15 μm (47 pF) detectors as a function of surface leakage and reverse current, as compared to noise form a 20 MΩ resistor in parallel with capacitors

A test of radiation damage was also carried out (see Fig. 12) with particularly indicative neutron fluences of the order of 1-10 M Rad, in comparison with a c-Si detector 255 μm thick.

This last one was unable to give any signal after 10^{13} n/cm^2, where a-Si: H detector was giving practically the same signal, but with a resolution worse by a factor of two. 10 Rads are a typical year dose in experimental apparatus positions, near the beam axis, for the next generation high energy accelerators.

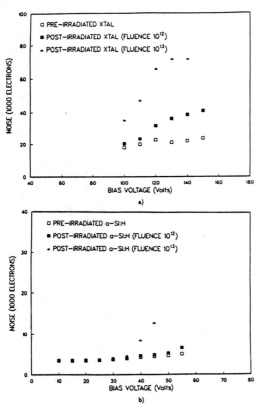

Fig. 12 —Detector noise (FWHM) before and after exposure to fast neutron fluences for (a) a crystalline Si detector and (b) an a-Si:H detector

Finally, electrons from a β⁻ source ($^{90}Sr-^{90}Y$) were used as minimum ionizing particles to test a-Si detectors.

Because the signal was below the noise, a trigger was used from an c-Si detector aligned to the source in coincidence with a-Si: H detector, to start an acquisition cycle in a digital sampling oscilloscope. An average is carried out every 256 pulses and the digitally stored result is sent to a computer that emulates a PHA giving an average energy spectrum. Fig. 13 gives the result of a β measurement for a 5 μm Schottky barrier detector: the centroid of β⁻ amplitude distribution

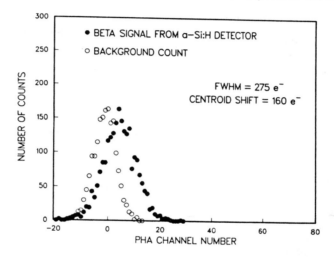

Fig. 13 - Pulse-height spectrum for an a-Si:H detector 5 μm thick on coincidence with β trigger from an aligned c-Si detector (°) and with a pulsar trigger (°). Each pulse is derived from a 256-event signal average

corresponds to a collection of 160 electron-hole pairs per electron transit, which means about 1 KeV as compared to 1.3 KeV expected theoretically (the 5 μm thickness is only nominal). This result is extremely good and proves that minimum ionizing particle detection could be achieved with a detector 16 times thicker (the noise adds in quadrature for the 256-trace average). A thicker detector, having lower capacity, would be expected to give lower noise.

Large area, low cost, particle detectors may be of some use for counting very weak activities of Th, U and Rn. Moreover, some use of a-Si could be made for smoke detectors and other apparatuses working with low alpha sources.

In heavy ion physics, a large area detector for angular distributions measurements, possibily equipped with a thin dE/dx detector in order to discriminate the ion mass by the product ExdE/dx, is certainly needed. In the case of low energies, such as Tandem accelerators and so on, the energy loss rate dE/dx could be in the range of several MeV/μm, and therefore, detectors thicker than 1 μm are certainly not needed. The best solution could be to deposit directly a thin p-i-n dE/dx detector on a c-Si detector, even if a combined ExdE/dx detector, with 1 mm strips for measuring angular distributions, realized with a-Si only, could be in the range of present technologies.

4. - MEDICINE AND X-RAY APPLICATIONS

X-ray detection from a Mo target (18 KeV) was successfully achieved[9] and used in order to determine the e-h pair creation energy. Pulses 2μs wide, filtered by 0.5 mm Al, were monitored by a c-Si detector and with different thin Si absorbers in order to calculate the expected detector signal as a function of depletion depth. Fig. 14 shows the ratio S_a/S_c of the signals from a 5 μm a-Si detector and from a c-Si detector of the same thickness.

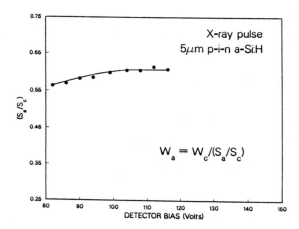

Fig. 14 - Amplifier output from X-rays pulses incident on a 5μm a-Si:H detector, in plateau conditions, normalized to the signal expected from a c-Si detector 5 μm thick placed on the same X-ray beam. Wa and WC are the respective radiation ionization energies for two materials

Since saturation is observed, the only explanation for the 60% signal ratio is the different values of W. The new value of W for a-Si matches Klein's curve by assuming Eg=1.9 ev for a-Si (see Fig. 8).

The first applications of a-Si to X-ray detection was carried out with different kind of sensors [6]. The former one was essentially a more or less standard p-i-n cell with the intrinsic region thickness of 0.9 μm in order to increase sensitivity. X rays were generated from an diffractometer fitted with a tungsten target.

Different sensor geometries were tried (see Fig. 15) with practically the same results. The sensitivity is not large, even if a good signal can be obtained from a 1 cm^2 area detector even at 10 R/min. The stability of this sensor is not good, however, since the output current decreases by 40% in 6 hours. For these reasons, a different approach

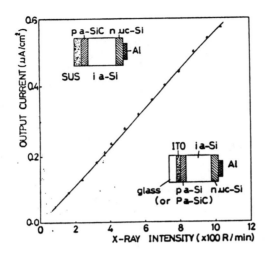

Fig. 15 - Typical structure and output characteristics of a-Si p-i-n X-ray photovoltaic sensor

has been tried, by depositing a layer of ZnS (Ni) powder phosphor on the front surface of glass substrate (see Fig. 16). Harder X-rays must be used in this case, since they should be incident on glass instead of on thin Al deposition. The visible green light excited by ZnS is very well matched to the spectral sensitivity of the a-Si pin cell (see Fig. 17). The sensitivity increases by a factor of 30 approximately, depending clearly on the phosphor thickness.

Moreover, the i region thickness can be modulated in order to get a better matching with other scintillators, like BGO or CdWO (see Fig. 18): this last one behaves very similarly to ZnS (see Fig. 15).

Also the stability of this kind of sensor is better: an illumination with 400 R/min for several hours with no change in output current was checked. In this case, since X rays are almost completely absorbed by

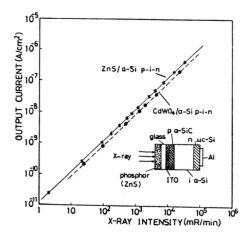

Fig.16—Typical structure and output characteristics of phosphor a-Si p-i-n detector (i region is 0.6 µm thick)

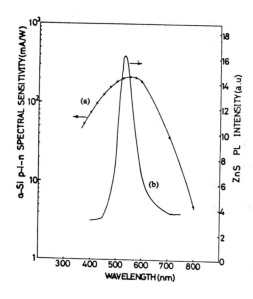

Fig.17—Spectral sensitivity of a-Si p-i-n photovoltaic cell and the fluorescence spectrum of ZnS phosphor film

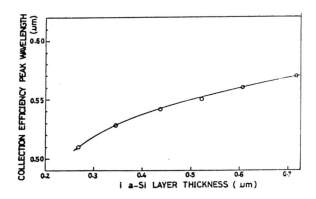

Fig. 18 - Peak wavelength of the collection efficiency as a function of the i-layer thickness in the p-i-n a-Si junction

the ZnS film, do not damage the cell.
In order to examine the possibility of a real application, a 16-channel CdWO4/a-Si p-i-n X-ray detector was made for a SCT-100 N type XCT medical equipment and image reconstruction tests were successfully performed.
Also photoconductive sensors were tried for this application: an example is shown in Fig. 19 for a sensor 1.5 μm thick with a gap arrangement of electrodes.
Sensitivity is much lower than the case of photovoltaic sensor, however the stability is very good.
In conclusion, no serious drawback is expected for application of a-Si detector to X-ray field.
In particular, addressable usage sensors suitably designed could be an interesting possibility for XCT medical equipment applications. One and

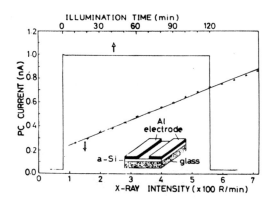

Fig.19 - Typical structure and output characteristics of photo-conductive a-Si sensor. The stability test has been performed with 400 R/min exposition rate

two-dimensional X-ray sensors may have good chances for both scientific and industrial applications.

Position sensitive X-ray detectors are needed in synchrotron radiation experiments[13] such as for instance, two-dimensional diffraction measurements or kinetic diffraction time-resolved experiments. Detectors should have a high efficiency, excellent position resolution, high count-rate capabilitities, good integral and differential linearity and a large variety of collecting areas. The best PSDs are certainly the gas proportional detectors which, however, are not as fast as required by the increased X-ray fluxes in present synchrotron sources.

The energy range of these experiments (3-20 KeV) makes in silicon PSDs quite interesting. However, leakage currents of ordinary silicon junction diodes are too large to obtain a good sensitivity for measurements carried out in the current mode (which is the only one allowed for high count rate experiments), even if cooled by liquid nitrogen. The amorphous diodes developped for sensing images are really much smaller and, moreover, the fabrication of amorphous silicon diodes arrays is easier than of other diode arrays.

Integrated photodiodes arrays with MOS multiplex switches were introduced earlier than amorphous ones, even if the increase in the dark current by X-ray irradiation on MOS switches was ramarkable.

Moreover, these integrated arrays are limited to a dimension of 3 cm, which is too short for using in synchrotron experiments.

Fuji-Xerox produces, for facsimile applications, a 2048 sensor, with pixel area 100 μm x 100 μm and a pitch of 125 μm. Each pixel is followed by a voltage amplifier wich reads the voltage of a charging 3pF capacitor. Multiplexers are also used to obtain a voltage pulse train and a reset for each capacitor is provided. Maximum readout speed is 1μs/ch. The detector response for an 8 KeV X-ray beam is shown in Fig. 20a. Evidently, the spatial resolution is limited by the pixel pitch or center-to-center spacing. The minimum detectable photon rate is determined by the fluctuation of the background pattern. The typical leakage current of each pixel is 0.2 pA, which is much less than 1pA of integrated photodiode arrays.

The dinamic range is therefore very large: 2000/1 at a measuring time of 1S. The estimated sensitivity could be 0.004 pA in the normal experimental conditions, which correspond to a photon rate of 12 ph/(s-ch) at 8 KeV. However, since the a-Si thickness is only 1μm, the absorption rate is 1.5% and the minimum detectable photon rate rises to 10^3 ph/(s-ch), which is really too low. Also in this case, in order to improve the sensitivity, a phosphor has been used (Gd_2O_2S) which has a peak wavelength at 544 nm, very well matched to the spectral response of the photodiode.

The conversion factor becomes eight times higher, but the resolution, for a phosphor thickness of 130 µm, rises to 375 µm (see Fig. 20b).

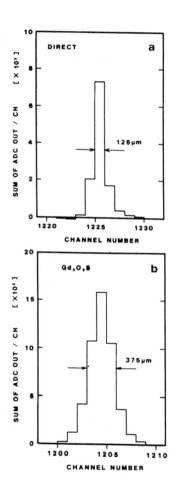

Fig. 20 - Spatial resolution of amorphous photodiode array: a) direct X-ray exposure; b) light-coupled X-ray PSD

Moreover, since the phosphor has a decay time of 1 ms, in time resolved experiments the time slice must be langer than 1 ms. In conclusion, for this kind of experiments, a detector with a much thicker sensitive layer is needed, a request that seems in the range of present technologies.

5. - CONCLUSIONS

Amorphous silicon detectors present several advantages with respect to c-Si ones for many applications: easiness of fabrication, low cost, adaptability to geometries, area, radiation hardrness, etc.
Noise and the relationaly slow rise times seem not to be a drawback for many applications.
For high energy physics applications, a particular detector, with at least 6 layers, 20 μm thick, should be developped in order to detect minimum ionizing particles with a S/N ratio of more than 6. Other alternatives are likely, particularly if the material quality, with respect to DOS values, will increase. For a good application in this field, DOS should drop to 10^{14} cm^{-3}, in order to be able to deplete thicker detectors (20 μm).
Other applications to calorimetry, CCDs and so on are suggested. Particularly pixel detectors should be considered, because of lower capacitance and noise, even if a fast trigger is not possible with a-Si and should be realized separately.
X-ray applications in medicine, industry and research are, on the contrary, a real actual possibility.
In general, for most applications, thicker sensitive regions are requested, as for c-Si CCDs.
The possible market dimensions are not easy to be defined, but the main

interest could be for medicine and also industrial X-ray applications. In any case, it is certainly a good challenge for an ever better quality a-Si:H.

References

1) P.G. LeComber et al, J. Non-Cryst. Solids 77&78 1373 (1985)
2) A.J.Snell, W.E.Spear, P.G. LeComber and K.D. Mackenzie, Appl. Phys. A26 83 (1981)
3) J.N.Kaplan, J.Morel, T.A.Mulera, V.Perez-Mendez, G.Schurnmacher and R.A. Street, IEEE Trans. Nucl.Sci. NS-33 351 (1986)
4) V.Perez-Mendez, S.N. Kaplan, W.Ward, S.Qureshi and R.A. Street, Nucl. Instr. Meth. A260 195(1987).
5) S.N.Kaplan, I.Fujeda, V.Perez-Mendez, S.Qureshi, W.Word and R.A. Street, London Conf. on Position Sensitive Detectors, Sept. 1987.
6) G.P.Wei, D.Kruangam, C.Y.Xu, H.Okamoto and Y.Hamakawa, 2nd Int. PVSEC, 1986 Beijng (China) p. 458.
7) K.Hasegawa, K.Mochiki, M.Koike, Y.Satow, H.Hashizume and Y. Iitaka, Nucl.Instr. Meth. in Phys. Res. A252 158 (1986)
8) D.Green, IEEE Tran.Nucl.Sci. NS-33 60 (1986)
9) V.Perez-Mendez, S.N. Kaplan, W.Word, S.Qureshi and R.A. Street, LBL-22756 (1987).
10) P.G.Rancoita and A.Sedman, Nucl.Instr.Meth. in Physics Res. A263 84 (1988)
11) V.Perez-Mendez, J.Morel, S.N.Kaplan and R.A.Street, Nucl.Inst.Meth in Physics.Res. A252 478 (1986)
12) R.A.Street, Phys.Rev.B27 27 (1983)
13) G.C. Smith, Nucl. Instr. Meth. 222 230 (1984)

AUGER AND ELECTRICAL STUDY OF THE TCO/Si INTERFACE IN AMORPHOUS SILICON DEVICES

G.Grillo, G.Conte, D.Della Sala, F.Galluzzi, V.Vittori

ENIRICERCHE S.p.A., Via Ramarini
00015 Monterotondo (Roma), Italy

ABSTRACT

The interface between the transparent conducting oxide (TCO) SnO_2 and amorphous silicon in optoelectronic devices (e.g. solar cells) may be damaged as a result of the interaction between the TCO and the plasma used for the glow-discharge deposition of the a-Si:H layers. Electrical barriers at the interface and diffusion of tin into the active layers may result from chemical reduction of SnO_2 and oxidation of Si.
In this study we use Auger depth profiling techniques to measure the TCO damage directly on the devices. A method is developed for quantifying the total amount of reduced tin from the profile data. Primary electron beam current densities much lower than previously used in the literature are used to avoid artifacts due to electron-beam-induced damage during the Auger measurements.
The extent of TCO reduction is correlated with the preparation procedures and with the photovoltaic performance of the cells. In particular, the beneficial role of thin protecting metal layers on TCO is investigated by both Auger and electrical measurements.

INTRODUCTION

In large-area thin-film optoelectronic devices based on amorphous silicon, such as photovoltaic cells, transparent conducting oxides (TCO) are generally used as front electrodes. High optical transmission, low sheet resistance, and good ohmic contact to the active amorphous layers are the prerequisite for the achievement of high device performances. Variously doped indium, tin and zinc oxides, or mixtures of them, are the most used TCO materials.
During the glow-discharge (GD) deposition of the amorphous Si layers, however, interaction of the TCO with the activated species (radicals, electrons, ions) present in the glow discharge plasma can lead to chemical reactions at the TCO/Si interface. Chemical reduction of SnO_2, oxidation of Si, interdiffusion at the interface, dopant loss, and diffusion of the metal from TCO into the active layers may result, with consequent rising of electrical barriers at the nominally ohmic contact and deterioration of the transport properties of the device. It is generally recognized that the quality of the TCO/amorphous Si interface is one of the important issues in the achievement of high performance in photovoltaic p-i-n cells, and this has brought about the substitution of In-Sn

oxide (ITO) by the supposedly less vulnerable tin oxide (TO) in the common practice of p-i-n preparation.

Severe damage, caused by prolonged exposure of TCO to glow-discharge plasmas, was detected by electrical (1) and optical (2) measurements; surface techniques (3) showed reduction of SnO_2 to bivalent and to metallic tin. It was found that the effects depend on the deposition temperature and on the chemical constitution of the plasma; hydrogen at temperatures greater than about 200 °C is particularly effective.

To investigate the chemistry of the actual SnO_2/Si interface, depth profiling techniques in conjunction with surface analytical spectroscopies such as X-ray Photoelectron Spectroscopy (XPS) and Auger Electron Spectroscopy (AES) can be, and have been, used (4-8). These studies were mainly intended to observe and understand, in purposely-made thin film structures, the mechanisms of the various phenomena (Sn reduction, Si oxidation, intermixing, diffusion of Sn into the active layers) occurring at the interface.

The aim of the present study is to develop a quantitative method for measuring "routinely" the TCO damage in actual devices, so that it can be used in assisting the device optimization.

EXPERIMENTAL

Commercially available polycrystalline fluorine-doped SnO_2 on glass was used as substrate (Nippon Sheet Glass, 800 to 4500 Å thickness). Amorphous hydrogenated Si layers and complete hetero-junction p-i-n photovoltaic cells were deposited in a home-made three-chamber glow-discharge apparatus equipped with load-lock introduction system, turbomolecular and rotary pumps, base pressure better than 10^{-5} Pa. The discharge was operated at 13.56 MHz, in the capacitively coupled mode. Pure and H_2-diluted silane, methane, H_2-diluted posphine and diborane were used as feedstock; deposition temperatures in the range 220-270 °C, RF powers from 2 to 7 watts, and pressures from 100 to 400 mTorr were used. The structure glass/ TCO/ p-SiC:H/ i-Si:H/ n-Si:H was adopted for the p-i-n solar cells. For the electrical measurement, an Al back-contact was deposited by sputtering on the cells. A few a-Si samples were also deposited by magnetron sputtering at room temperature.

For the electron-excited Auger spectra and depth profiles, a Varian scanning spectrometer was used in the derivative mode. The coaxial electron gun was operated at 2 keV, with electron beam currents varying from .5 to 4 µA. A Ne+ scanning ion gun, operated at 5 keV, was used for ion milling. The spectrometer was driven by a home-built data acquisition system, based on a IBM PC. The profiles were obtained by measuring cyclically the Auger signal at selected energies during ion milling.

During the measurements, the primary electron beam density was kept low (in the range of .1 mA/cm^2) by scanning the beam position at TV speed over a .6 x .6 mm^2 area. The ion gun also was rastered, over a larger area (.8 x .8 mm^2), to ensure a flat ion-milled crater. In these conditions, depth resolution is limited by the asperities developed during ion milling, and depends on the crystallinity and surface morphology of TCO. For this reason, only flat (i.e. non-textured) TCO was investigated; nevertheless, the depth resolution was found to depend on TCO thickness, and ranged from 200 to 500 Å.

A chemical shift of about 6 eV in the position of the Auger MNN lines of tin is observed upon oxidation (9). In the derivative mode, the two main peaks at 430 and 437 eV in metallic tin shift to 424 and 430 eV respectively in SnO_2. In spite of the overlap of the two spectra in the 430 eV region, the presence of elemental tin at the interface was easily detected on the 437 eV line in the case of severely damaged TCO. In most of the samples studied, however, the effect was very small, and manifested itself only as a shoulder in the 430 eV line, so that a sensitive method had to be found to detect and quantify the presence of reduced tin from the relatively few spectral points obtainable in the depth profiles.

To this aim, the following procedure was adopted: the spectra of the two Sn species were schematized as though they were composed of two lines each, with intensity coefficients given by i_{ol}, i_{oh}, i_{rl}, i_{rh} for the low- and high-energy peaks of SnO_2 and Sn, respectively. If the atomic concentration of oxidized and reduced tin are C_o and C_r, we have, at any depth z

$$I_{424}(z) = i_{ol} C_o(z)$$
$$I_{430}(z) = i_{oh} C_o(z) + i_{rl} C_r(z)$$
$$I_{437}(z) = i_{rh} C_r(z) .$$

Rather than try to resolve the two overlapping spectra, we divide the total spectrum into a low- and a high-energy part, with intensities I_l and I_h:

$$I_l(z) = I_{424} = i_{ol} C_o(z)$$
$$I_h(z) = I_{430} + I_{437} = i_{oh} C_o(z) + (i_{rl}+i_{rh}) C_r(z).$$

Well in the bulk of the tin oxide layer, no metallic tin is supposed to be present, so that $I_l(\infty) = i_{ol}$ and $I_h(\infty) = i_{oh}$. We can thus use these two values to normalize the measured intensities, and get

[1] $$I_h(z)/I_h(\infty) - I_l(z)/I_l(\infty) =$$
$$= C_r(z) (i_{rl}+i_{rh})/i_{oh} \simeq 2 C_r(z),$$

where, for the last equation, the intensity coefficients of the four lines are supposed to be about equal.

The procedure was implemented by an off-line elaboration on the stored profile data and is illustrated in fig. 1. At each z, the sum of the points measured in the 422-428 and 428-438 intervals was taken as I_l and I_h, respectively. The final result is a profile of the concentration, C_r, according to eq. [1], of metallic Sn present at the interface; since, however, in our measurements the width of the interface was presumably narrower than the obtainable depth resolution, we will consider as meaningful only the integral of this concentration

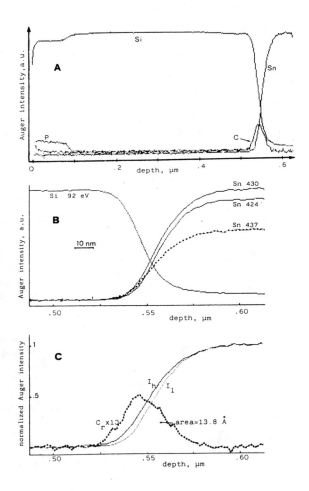

Fig. 1. A) Auger depth profile of a complete p-i-n solar cell.
B) Enlargement of the interface region.
C) Quantification of elemental Sn, according to eq. [1].

over the interface, which gives the total quantity of elemental tin present, expressed in depth units.

The above method is not claimed to give a very accurate absolute value, but has the merit of giving exactly zero if there is no change in the shape of the spectrum during the profile (as it should be when only SnO_2 is present), and of being independent of the instrumental settings and of any previous knowledge of the sensitivity factors.

Besides SnO_2 reduction, Si oxidation is also observed at the interface by Auger, and could possibly be taken as a significant index of interface damage. The Si oxide Auger line, however, is difficult to quantify by an "automatic" algorithm, because it is broad and weak, and is located in a spectral region with rapidly varying background close to the much more intense line of elemental Si. For this reason the Si lines were not considered in the present study.

RESULTS AND DISCUSSION

Table I reports the thickness of elemental Sn measured by Auger depth profiling at the a-Si/SnO_2 interface in a series of photovoltaic cells and in a few thin a-Si films deposited on TCO.

A strong dependence of the apparent damage on the current density of the primary electron beam used for the Auger measurements was found: measurements # 4 to 7 refer to the same sample, but were performed using different electron current densities. The conditions of measurement # 7, in which .4 A/cm^2 at 2 keV were used, are "routine" conditions often used in medium-resolution scanning Auger spectrometers, and are actually comparable to that used in previous literature measurements (for example in (8), while in (7) .01 A/cm^2 at 5 keV were used, and most of the other authors do not report the conditions). From comparison with the measurements # 4-6 on the same sample, it is apparent that, with this "routine" electron current density, the measured damage is mostly an artifact due to electron-beam-induced reduction of SnO_2; no meaningful or reproducible results could be obtained in measurements performed under these conditions.

The behaviour of SnO_2 under electron bombardment is somewhat deceptive, in that no change is seen in the Auger lineshape on free TCO surfaces upon prolonged bombardment. The change is, instead, readily seen in the presence of potentially reducing species, in our case Si; that is, SnO_2 does not give up its oxygens to vacuum but only to more oxidable atoms. We observed this by taking Auger spectra in sample regions (for example at the borders of the a-Si:H film) in which TCO was covered by only a few atomic layers of Si, and finding that, under "routine" electron bombardment, the shape of the Sn doublet changed with time to that of metallic tin. This, on the other hand, is consistent with the reports that TCO is reduced much faster under a hydrogen plasma than under an Ar plasma (1,2).

The conditions of measurements # 4 and 5, instead, (low current, rastered electron beam, current density about .2

mA/cm^2, 2 keV energy) were found to be free from electron-induced damage, and were adopted as the "safe" conditions for the other measurements.

One should be concerned also with ion-bombardment-induced damage. The total ion dose per cm^2 necessary for ion milling through the interface is approximately fixed, so one cannot easily compare damages, if present, induced by different ion intensities. In the present study, however, ion currents of less than .1 mA/cm^2 were used, and, generally speaking, one would not expect ions to be more effective than electrons in causing chemical reduction. In order to check this assumption, TCO reduction was measured on control a-Si samples deposited on TCO by magnetron sputtering at room temperature; as expected, little or no damage was found (meas. # 1-3), and this supports the substantial correctness of the whole procedure.

The data of tab. I which were taken under the "safe" conditions are reported versus the deposition temperature in fig. 2. Despite some scatter in these data, the correlation of the interface damage with the deposition conditions can be clearly seen.

A damage of 12-15 Å is measured on photovoltaic cells deposited under our typical conditions (235 °C, 4-5 watt RF power for the deposition of the first B-doped a-SiC:H); an increase of the damage with increasing deposition temperature and RF power is found; the highest damage occurs in a cell (corresponding to # 13 in tab I) in which both the deposition temperature and the RF power were high (260 °C and 7 W); the lowest damage is found on a cell (# 12) in which the deposition of the first layer was started at low temperature, so to create a thin protective "ceramic" SiC layer on the TCO substrate. Dependence on the deposition geometry is shown by the measurements # 8-9 on a a-Si:H film prepared in a triode apparatus, in which contact between the glow discharge and the substrate was prevented by a grounded electrode.

The measurement method was also applied to cells in which the TCO substrate had been "protected" by previous deposition of thin (10 to 50 Å) Pt films. Table II shows the comparison between protected and unprotected samples deposited under the same conditions. It is expected that the thin metal film acts as a barrier for the reaction between the reducing species and TCO (10). The Auger results seems not always to confirm this. A damage decrease is seen in two Pt-protected samples, while a pronounced increase is seen in the third one: this can be ascribed, however, to a spectral interference between the lines of SnO_2 and Pt, which has a broad Auger line at 445 eV. The presence of this line can alter the delicate balance on which the quantifying algorithm is based, and in fact in the sample of meas. # 29 there was a relatively thick (about 50 Å) Pt film; so it has to be concluded that the "automatic" method is unreliable in the presence of thick Pt films at the interface, even though visual inspection of the spectra showed qualitatively that a distinct decrease in elemental tin was indeed obtained in Pt-protected cells.

Photovoltaic performance enhancements were consistently found in cells deposited on metal-protected TCO. The i-V characteristics taken under simulated AM1.5 illumination show that

Tab. I

#	sample	SnO$_2$ damage, Å	depos. temp., °C	a-Si thick., Å
1	sput'd 1	-2.2	50	700
2	"	0.0 *	-	-
3	sput'd 2 §	4.1	50	900
4	i triode	9.0	240	3400
5	"	10.2	-	-
6	"	11.8 *	-	-
7	"	40.0 * £	-	-
8	i diode	7.2	220	3700
9	"	10.2	-	-
10	p-SiC	10.8	235	550
11	cell 24 &	11.9	235	5600
12	cell 48	8.2	235	5250
13	cell 54	20.1	100	3100
14	cell 56	12.4	260	4100
15	cell 57	14.9	235	5200
16	cell 61	12.8	235	6100
17	cell 62	9.0	210	5450
18	cell 63	14.4	210	5300
19	cell 64	11.6	235	6450
20	cell 65	13.8 *	235	5800
21	cell 86 §	15.7	270	5650
22	" §	21.2 *	-	-
23	cell 95 §	13.8	235	3600

- Measuring conditions:
  ion beam: 5 keV Ne$^+$, 60 to 80 Å/min
  electron beam: 2 keV, .7 µA scanned over .6x.6 mm^2
  except: £ beam focused to 30 µm, no scanning
  * current 3 µA
- Samples:
  # 1-3 magnetron-sputtered a-Si on TCO
  # 4-9 GD-deposited a-Si:H on TCO
  # 10 GD-deposited B-doped a-SiC:H on TCO
  # 11-23 p-i-n photovoltaic cells on TCO
- TCO thickness 800 Å, except: & 2000 Å, § 4500 Å

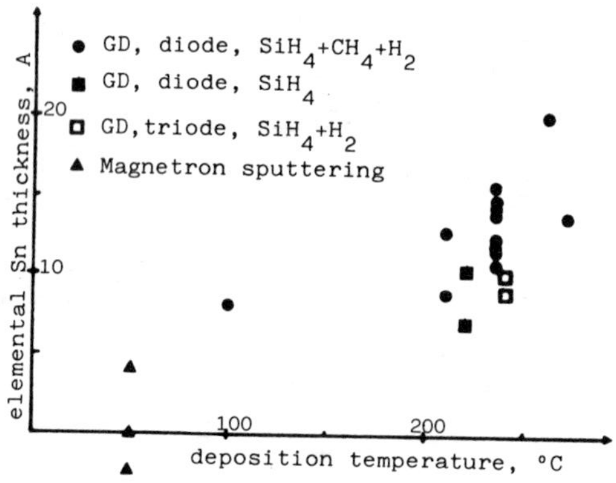

Fig. 2

● GD, diode, SiH$_4$+CH$_4$+H$_2$
■ GD, diode, SiH$_4$
□ GD, triode, SiH$_4$+H$_2$
▲ Magnetron sputtering

elemental Sn thickness, Å vs deposition temperature, °C

Tab. II

#	sample	SnO$_2$ damage, Å	depos. temp., °C	a-Si thick., Å
24	cell 54	20.1	260	3150
25	cell 54/Pt	18.6	-	-
26	cell 95 §	13.8	235	3600
27	cell 95/Pt §	8.6	-	-
28	p-SiC	10.8	235	550
29	p-SiC / Pt	28.6	-	-

- Measuring conditions: as in tab I
- Samples:
  # 24-29 p-i-n photovoltaic cells on TCO
  # 30-31 GD-deposited B-doped a-SiC:H on TCO
- TCO thickness: as in tab. I

fill factor improves (11), due mainly to a decrease in series resistance; a 73% fill factor value was obtained in a cell whose TCO had been previously coated by evaporated Pd, with respect to 68% measured in the same cell deposited directly on TCO. The same trend was observed also with evaporated Pt.

CONCLUSIONS

A method for quantifying the damage at the TCO/a-Si interface in amorphous-silicon-based devices was developed and was applied to actual photovoltaic cells, as well as to purposely-made thin film structures. The method is based on a detailed analysis of Auger depth-profile data for quantifying the chemical reduction of SnO_2 at the interface. A careful choice of the measurement conditions was found to be the essential prerequisite to avoid artifacts due to electron-beam-induced reactions.

The method works well on devices deposited directly on TCO; in typical working solar cells 12 to 15 Å of elemental tin was measured at the interface; the damage was seen to increase with increasing temperature and RF power. In devices in which TCO had been previously coated by a thin protective Pt layer, substantial decrease of damage was seen qualitatively on the spectra, but spectral interferences prevented correct use of the quantitation algorithm.

The electrical measurement showed performance enhancements in the j-V characteristics of the cells deposited on TCO coated by a thin Pt or Pd film; lower series resistance and higher fill factor values, up to 73%, were obtained.

Work supported by AGIP S.p.A.

REFERENCES

(1) O.Kuboi, Japan J. Appl. Phys., vol. 20, p. 1783, 1981
(2) H.Schade, Z.E.Smith, J.H.Thomas III, and A.Catalano, Thin Solid Films, vol. 117, p.149, 1984
(3) J.H.Thomas III, Appl. Phys. Lett., vol.42, p.794, 1984
(4) J.H.Thomas III and A.Catalano, Appl. Phys. Lett., vol.43, p.101, 1983
(5) M.Kitagawa, K.Mori, S.Ishigawa, M.Ohno,Y.Yoshioka, and S.Kohiki, J. Appl. Phys., vol. 54, 1983
(6) S.Badrirayanan, S.Sinha, and A.P.B.Sinha, Thin Solid Films, vol. 144, p. 133, 1986
(7) E.Esen and F.Ramos, Proc. 7th E.C. Photovoltaic Solar Energy Conf., Reidel Publ. Co., Dordrecht 1987, p.560
(8) A.S.Verlinde, J.Smeets, P.Nagels, H.J.van Daal, and H.H.Bronsgerma, Proc. 8th European Photovoltaic Solar Energy Conf., Kluwer Academic Publ., Dordrecht 1988, p.791
(9) R.A.Powell, Appl. Surface Sci., vol. 2, p. 397, 1979
(10) K.Seki, H.Yamamoto, A.Sasano, and T.Tsukuda, Appl. Phys. Lett, vol. 44, p.682, 1984
(11) G.Conte, D.Della Sala, F.Galluzzi, G.Gramaccioni, G. Grillo, R.Tomaciello, and V.Vittori, Proc. 8th E.C. Photovoltaic Solar Energy Conf., Kluwer Acad. Publ., Dordrecht 1988, p.1647